Sharks - Past, Present and Future

Edited by Mohamed Nejmeddine Bradai, Bechir Saidi and Samira Enajjar

Published in London, United Kingdom

Sharks - Past, Present and Future
http://dx.doi.org/10.5772/intechopen.102161
Edited by Mohamed Nejmeddine Bradai, Bechir Saidi and Samira Enajjar

Contributors
Samuel K. Amponsah, Rachel Ackah, William Dzisenu Amekor, Asiedu Berchie, Andrews Apraku, Sondes Marouani, Sami Karaa, Othman Jarboui, Mohamed Nejmeddine Bradai, Bechir Saidi, Samira Enajjar, Hasan Alkusairy, Adib Ali Saad

First published in London, United Kingdom, 2023 by IntechOpen
IntechOpen is the global imprint of INTECHOPEN LIMITED, registered in England and Wales, registration number: 11086078, 5 Princes Gate Court, London, SW7 2QJ, United Kingdom

British Library Cataloguing-in-Publication Data
A catalogue record for this book is available from the British Library

Additional hard and PDF copies can be obtained from orders@intechopen.com

Sharks - Past, Present and Future
Edited by Mohamed Nejmeddine Bradai, Bechir Saidi and Samira Enajjar
p. cm.
Print ISBN 978-1-83768-313-0
Online ISBN 978-1-83768-314-7
eBook (PDF) ISBN 978-1-83768-315-4

Meet the editors

Mohamed Nejmeddine Bradai is a Professor of Higher Education at the National Institute of Marine Science and Technologies and head of the Marine Biodiversity Laboratory in Tunisia. He obtained an engineering degree in marine living resource exploitation as well as a master's degree, Ph.D., and a state doctorate degree. He is a member of the Shark Specialist Group of the International Union for Conservation of Nature (IUCN) and former coordinator of the subcommittee "Marine Environment and Ecosystems" of the Scientific Advisory Committee (SAC) of the General Fisheries Commission for the Mediterranean of the Food and Agriculture Organization (GFCM-FAO). He is a contributor to the Mediterranean Large Elasmobranchs Monitoring (MEDLEM) database. As a consultant on elasmobranchs of the United Nations Environment Programme's Regional Activity Centre for Specially Protected Areas (SPA/RAC), Dr. Bradai elaborated the Action Plan for the Conservation of Cartilaginous Fishes in the Mediterranean Sea (2020) and the National Action Plan for the Conservation of Cartilaginous Fishes (chondrichthyans) in the Mediterranean Coasts of Egypt (2021).

Bechir Saidi is a marine biologist who obtained his Ph.D. from the University of Sfax, Tunisia, on the taxonomy, biology, and fishery studies of sharks. Since 2002, he has been a member of the Marine Biodiversity Laboratory at the National Institute of Marine Sciences and Technologies (INSTM), Tunisia, where he conducts research on elasmobranchs to assess the impact of anthropic-related activities on their conservation. He has been involved in several national and international projects dealing with their biology, taxonomy, and fisheries interactions with the aim of producing comprehensive data to promote conservation action. Dr. Saidi is a member of the Shark Specialist Group of the International Union for Conservation of Nature (IUCN).

Samira Enajjar is a marine biologist in the Marine Biodiversity Laboratory of the National Institute of Marine Sciences and Technologies (INSTM), Tunisia. Since 2000, she has been conducting research on biodiversity, fishery, and the taxonomy of elasmobranchs, mainly in batoids in the Gulf of Gabès, Southern Tunisia. She obtained her Ph.D. with work on the Rhinobatids family. Dr. Enaijar contributed to the creation of vulnerable marine vertebrate training manuals. She is also a member of the Med Bycatch Project as a supervisor and expert on elasmobranchs. She is regularly engaged in training courses related to the taxonomy of rays in many national and regional projects. She collaborated on the development of the Mediterranean Angel Sharks: Regional Action Plan and is a member of the Food and Agriculture Organization's MedSudMed working groups on elasmobranchs of the South-Central Mediterranean.

Contents

Preface

Cartilaginous fishes or Chondrichthyans are an ancient group that appeared around 500 million years, even before the appearance of bony fish. Their current representatives are sharks, rays, and chimeras, hereafter referred to as 'sharks'. Some primitive forms of sharks are still present. These animals have been able to survive and overcome five mass extinctions since their initial appearance. They play a key role in maintaining the balance of marine ecosystems.

Nowadays, sharks are facing a worldwide extinction crisis; since 1970, the global abundance of oceanic sharks and rays has declined by 71%, owing to an 18-fold increase in relative fishing pressure. This depletion has increased the global extinction risk to the point at which three-quarters of the species comprising this functionally important assemblage are threatened with extinction. The International Union for the Conservation of Nature (IUCN), recognized at the international level to provide scientific advice on the conservation status of living species, confirm through its "Red List of Threatened Species" that a high percentage of sharks are threatened and many species are data deficient (i.e., their biology and ecology are unknown). Sharks face many challenges, such as habitat losses, pollution, climate changes, and so on, but fishing pressure seems to be the most critical.

Because the disappearance of sharks threatens the oceans and humanity, scientists, decision-makers, NGOs, and other stakeholders require improved knowledge of shark species and how to protect them. This book presents data on sharks' biology and ecology and discusses the impact of threats and how to reduce them. It is designed to bring awareness to the plight of sharks and their conservation. It is divided into three sections on "Ecology and Life History," "Sharks Fisheries," and "Management and Conservation Options."

Chapters 1 and 2 provide information on shark ecology, life history, and taxonomy. Chapters 3 and 4 describe shark fisheries (fishing gear, landing statistics, etc.) and their negative impact on shark populations. Chapter 5 focuses on the current status of cartilaginous fishes in the Mediterranean Sea, which is similar to other marine areas, and discusses progress on conservation measures and actions taken mainly through regional plans by regional fisheries management organizations (RFMOs) and nongovernmental organizations (NGOs).

Since the future of marine biodiversity and sustainable fishery majorly depends on elasmobranchs, more successful conservation of this fish group should be ensured, and urgent awareness is needed focusing mainly on the following priorities:

- Increase training and awareness of fishermen and controllers of fisheries on protected species and develop an effective control of fisheries

- Improve shark conservation by:

    reducing bycatch mainly through species release

    mapping and monitoring of critical habitats

- Improve collection of elasmobranchs landing statistics
- Strongly ban fishing and landing of vulnerable species
- Improve studies on stock assessment

To develop this research field, experts should focus on (1) biological parameters, (2) knowledge of fisheries, (3) valuable and available statistics, (4) taxonomy and shared stocks, (5) studies on migration and exchange between populations, and (6) choosing better evaluation methods.

I would like to thank all the chapter authors for their valuable contributions. I am also grateful to my two co-editors, both of whom are experts on sharks and rays. Finally, I wish to acknowledge the staff at IntechOpen for their help throughout the editorial process.

Dr. Mohamed Nejmeddine Bradai and Samira Enajjar
National Institute of Science and Technology of the Sea,
Marine Biodiversity Laboratory,
Carthage University,
Carthage, Tunisia

Bechir Saidi
Faculty of Sciences and Techniques of Sidi Bouzid,
Kairouan University,
Kairouan, Tunis

Section 1

Ecology and Life History

Chapter 1

Elasmobranchs in Tunisia: Status, Ecology, and Biology

Samira Enajjar, Bechir Saidi and Mohamed Nejmeddine Bradai

Abstract

The authors have compiled published information on taxonomy, distribution, status, statistics, fisheries, bycatch, biologic, and ecologic parameters mainly on food and feeding habits and reproductive biology of elasmobranchs along the Tunisian coasts. This bibliographic analysis shows that cartilaginous species, including sharks and rays are by far the most endangered group of marine fish, with 63 species, about 53% of all are critically endangered, endangered, or vulnerable. Overfishing, fishing practices, and habitat degradation are leading to dramatic declines of these species. Biologic parameters concern a few species primarily in the Gulf of Gabes. Therefore, recommendations to fill gaps in order to protect and manage elasmobranchs stocks are proposed in this chapter.

Keywords: status, elasmobranchs, Tunisia, bibliographic analysis

1. Introduction

The Mediterranean Sea is known to be an important habitat for elasmobranchs with at least 48 sharks and 38 batoids species [1]. However, the region is a hotspot of extinction risk [2]. It has been demonstrated that sharks in the Mediterranean Sea have declined by more than 97% in number and "catch weight" over the last 200 years [3]. This situation driven a regional and a global rising concern about shark conservation and management [4].

Tunisian coasts (Central Mediterranean Sea) are characterized by their sharks and rays diversity [5, 6]. The region is known to be an important habitat for this group and a breeding grounds for many species such as the sandbar shark (*Carcharhinus plumbeus*) [7–9]. Like the rest of the Mediterranean, Elasmobranchs in Tunisia are subject to an increasing pressure due to the anthropogenic activities mainly fisheries [10]. The emerging picture illustrates a decline of several elasmobranch populations [11]. Nevertheless, investigation on management and conservation on elasmobranch have received little attention [12].

Elasmobranchs are vulnerable to fishing mortality owing to their life histories characteristics, such as low fecundity, late maturity, and slow growth rates [1]. Accordingly, information on biology, ecology, fishery, distribution, and population structure is required for suitable management and conservation of this group. Unfortunately, the investigations research related to these creatures is quite recent, it started by the end of the 1990s when landings declined, and some species became threatened [6].

Along Tunisian coasts, research interested on elasmobranch has started in early 1970. Although the studies relating to this group of fish are maintained until today, several gaps still exist for an adequate management of catches and sustainable conservation.

The aim of the present chapter is to review and analyze the research publications relating to elasmobranch species along the Tunisian coasts in order to appreciate the main gathered information and gaps. In addition, this analysis will guide our future research in order to acquire the essential information indispensable for an adequate conservation of this group.

2. Study area

Tunisia, with 2290 km of coastline, constitutes a transition zone between the eastern and western basins of the Mediterranean [13]. The Tunisian marine coasts include the FAO-GFCM Geographical Sub Areas (GSA) 12, 13, and 14 (**Figure 1**).

The northern coasts (GSA 12) are characterized by a turbulent underwater morphology, an alternation of hard, and soft bottom and a steeply sloping continental shelf. This diversity of biotopes gives them a high biodiversity. Among the 327 fish species listed in Tunisian waters, 270 were recorded in the Northern coast [5].

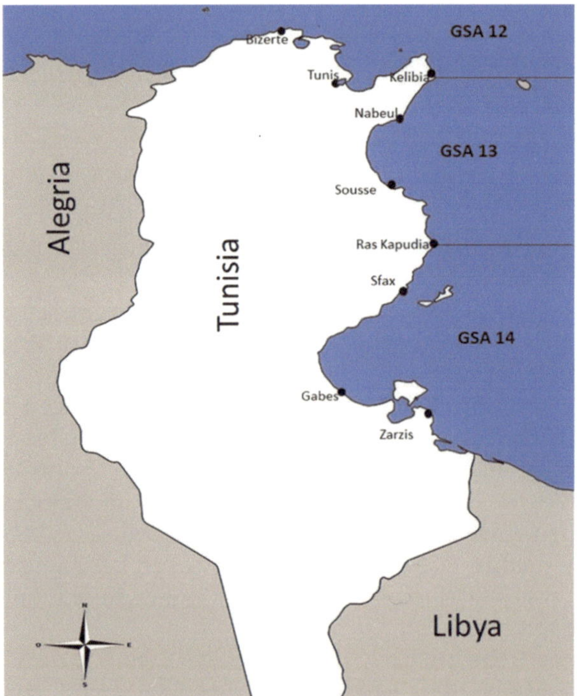

Figure 1.
GFCM geographic subareas off Tunisian coasts.

The eastern region of Tunisia (GSA 13), corresponding to the Gulf of Hammamet, begins with a narrow continental shelf (the -50 m isobath is located quite far from the coast), bordered by the Siculo-Tunisian channel and gradually widening from north to south of this region. The seabed of the area provides a transition between the northern and southern Tunisia [14, 15].

The GSA14, corresponding sensu lato to the Gulf of Gabes, represents the southern part of the Tunisian coast [16]. This region is characterized by a significant tidal phenomenon and an extended continental shelf. The presence of extensive seagrass meadows and the ease of access to fishing areas rich in species of high commercial value makes this region one of the most important maritime fishing areas in Tunisia.

The area is a high spot for marine biodiversity of regional importance. It constitutes a preferential habitat for several emblematic vertebrates: a wintering and feeding area for the Loggerhead Sea Turtle (*Caretta caretta*) [17], a nursery for several species of elasmobranchs, some of which are threatened [9–12], and a favorable area to several fish such as the groupers and tunas. Cetaceans, especially bottlenose dolphin (*Tursiops truncatus*) and the fin whale (*Balaenoptera physalus*), are regularly encountered [18, 19].

3. Elasmobranchs landing

In Tunisia, the elasmobranchs species are caught as bycatch. Nevertheless, some species such as the sandbar shark and the smooth hound are targeted by a small artisanal fishery in the southern coast of the country during the summer [20]. This fishery uses a specific gill nets locally called "Garracia" and "Gattatia".

Elasmobranchs represent an average of 2% of the national landing [21]. According to FAO Statistic, a mean of 2370 tons' year is landed during the last 20 years (2000–2020). The production shows an increasing trend, although some exceptional decrease is noted during 2012 and 2017 (**Figure 2**).

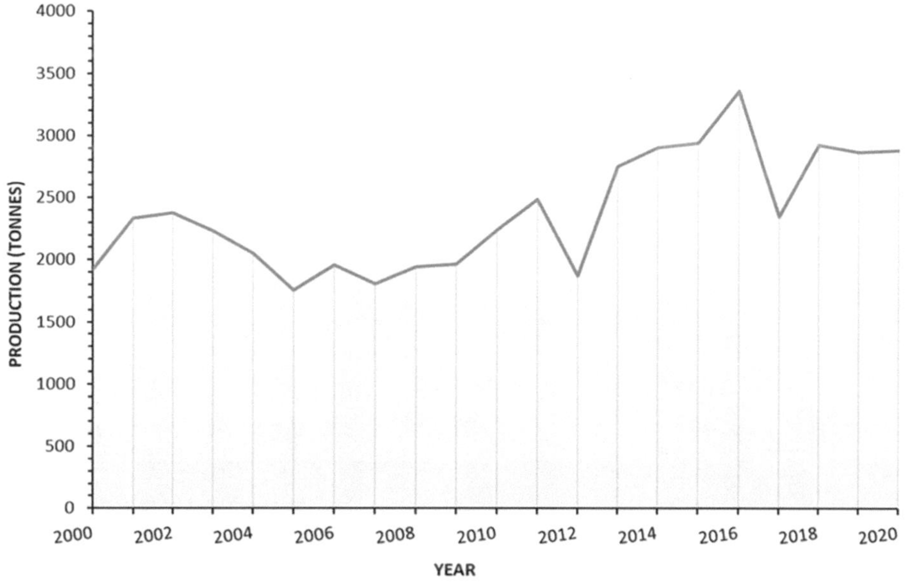

Figure 2.
Tunisian elasmobranchs production according to FAO statistics from 2000 to 2020.

The Gulf of Gabès region (GSA 14) is known to be the most important area for sharks and rays in Tunisia, contributing by more than 60% in the landing of elasmobranchs [10]. However, during the last years, the statistics data provided by the General Directorate for Fisheries and Aquaculture (GDFA) between 2008 and 2020 show an increase in landing of elasmobranch of the Eastern region (GSA 13) (**Figure 3**). This area contributed in 2020 by more than 49% in the Elasmobranchs national production.

Along the Tunisian costs, elasmobranchs are landed mainly by small-scale vessel using gillnets, trammel nets, and longlines followed by bottom trawl (**Figure 4**).

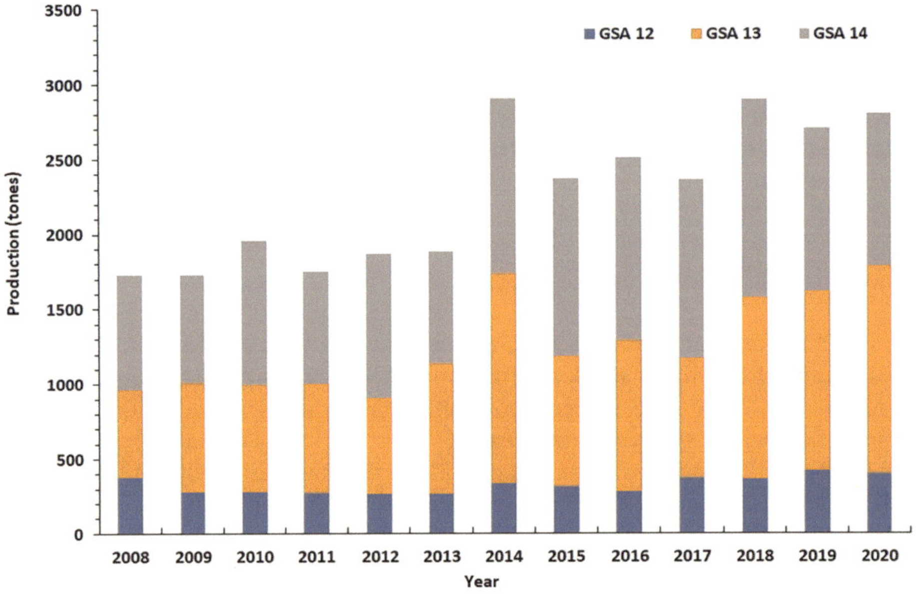

Figure 3.
Tunisian elasmobranchs production by GSA according to GDFA statistics from 2008 to 2020.

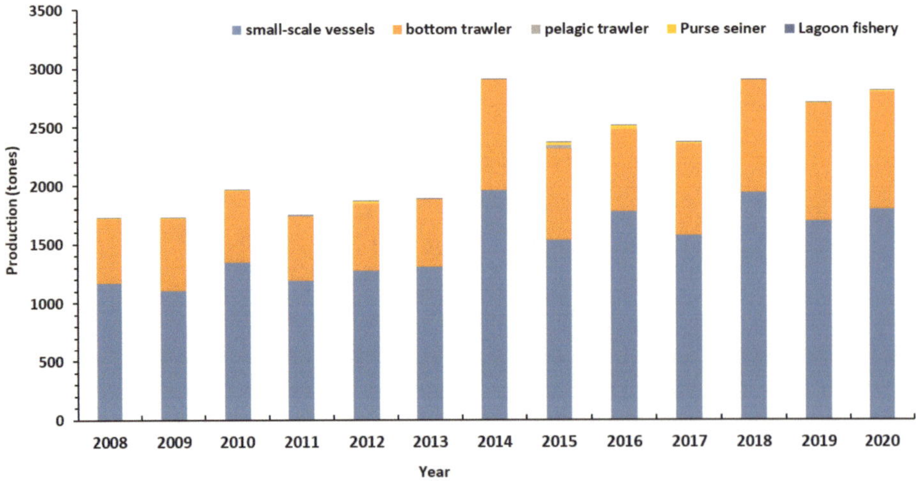

Figure 4.
Tunisian elasmobranchs production by fishing gear according to GDFA statistics from 2008 to 2020.

4. Diversity and status

The list of elasmobranchs species occurring in Tunisian waters is established mainly by the authors and bibliographic analysis over the last 20 years. Species are classified according to four categories: very common, common, rare, and very rare. The analysis of data shows the occurrence of 63 elasmobranchs species in the area: 37 sharks belonging to 17 families and 26 batoids belonging to eight families (**Table 1**). This number reflected the specific richness on species in the area (71.6% of species signaled in the Mediterranean Sea).

Four species cited in the literature are not considered in this list because their record seems to be doubtful or not observed during the study period: the cuckoo ray (*Leucoraja naevus*), the African ray (*Raja Africana*), the spiny dogfish (*Squalus acanthias*), and the porbeagle (*Lamna nasus*). However, three species were observed for the first time in the area during the study period: the shortnose spurdog (*Squalus megalops*) [22], the little sleeper shark (*Somniosus rostratus*) [23], and the bigeye thresher (*Alopias superciliosus*) [24].

The spinetail devil rays (*Mobula japonica*) signaled in the area in 2015 [25] are not considered in this list because it was assessed by many authors as a junior synonym of the devil fish (*Mobula moblar*) [26, 27]. No proofs to support the hypothesis of two different species were demonstrated.

Among elasmobranchs species occurring in Tunisian coast, only three species were very common in all sub-area; species caught very frequently throughout the region along the year; the smooth hound (*Mustelus mustelus*), the common torpedo (*Torpedo torpedo*), and the thornback ray (*Raja clavata*). Five species were common; species captured in more or less abundant quantities in at least one sector of the region and during a period of the year; the shortfin mako (*Isurus oxyrinchus*), the marbled electric ray (*Torpedo marmorata*), the brown ray (*Raja miraletus*), the round stingra (*Taeniura grabatus*) and the blackspotted smooth-hound (*Mustelus punctulatus*).

The southern waters of Tunisia were characterized by the presence of costal species: the blackchin guitarfish (*Glaucostegus cemiculus),* the common guitarfish (*Rhinobatos rhinobatos*), and the spiny butterfly ray (*Gymnura altavela)*, whereas deep species were found mainly in northern zone: the Velvet belly (*Etmopterus spinax*), the kitefin shark (*Dalatias licha*), and the little gulper shark (C*entrophorus cf. uyato).*

The number of species recorded in each GSA are almost comparable: 51 species in GSA 12 and 14 and 52 in GSA 13.

According to IUCN red list, more than 52% of elasmobranch species observed in Tunisian waters were threatened (Critically endangered, endangered, and threatened). Thirteen species were data deficient, not evaluated, or not applicable (**Figures** 5 and **6**).

5. Available data on elasmobranchs (Bibliographic analysis)

Two hundred and fifty-four references concerning elasmobranch species off Tunisia were published between 1971 and 2022. The temporal distribution of publications indicated that attention on elasmobranch has started in 1970. However, there is a lack of studies in the area during the period from 1980 to 1999. Since 2000, an interest in research on elasmobranchs is noticed in the area (**Figure** 7), following the emergence of an international concern for the conservation of this group of fish. However, studies concern mainly species of the southern (GSA 14) and northern coasts of the

	GFCM Sub area			IUCN Status
Species	**GSA 12**	**GSA 13**	**GSA 14**	
Hexanchiformes				
Hexanchidae				
Heptranchias perlo (Bonnaterre, 1788)	R	R	R	DD **
Hexanchus griseus (Bonnaterre, 1788)	R	C	R	LC
Lamniformes				
Lamnidae				
Carcharodon carcharias (Linnaeus, 1758)	R	R	R	CR *
Isurus oxyrinchus Rafinesque, 1810	C	C	C	CR *
Odontaspididae				
Odontaspis ferox (Risso, 1810)		TR		CR *
Carcharias taurus Rafinesque, 1810			TR	CR *
Alopiidae				
Alopias vulpinus (Bonnaterre, 1788)	R	R	R	EN **
Alopias superciliosus Lowe, 1841		VR		EN
Cetorhinidae				
Cetorhinus maximus (Gunnerus, 1765)	VR	VR	VR	EN **
Carcharhiniformes				
Pentanchidae				
Galeus melastomus Rafinesque, 1810	C	C	R	LC
Scyliorhinidae				
Scyliorhinus canicula (Linnaeus, 1758)	VC	VC	C	LC
Scyliorhinus stellaris (Linnaeus, 1758)	R	R	VR	NT
Triakidae				
Mustelus asterias Cloquet, 1821	R	R	R	VU **
Mustelus mustelus (Linnaeus, 1758)	VC	VC	VC	VU **
Mustelus punctulatus Risso, 1827	C	C	C	VU **
Galeorhinus galeus (Linnaeus, 1758)	R	R	R	VU *
Carcharhinidae				
Carcharhinus plumbeus (Nardo, 1827)	R	C	VC	EN **
Carcharhinus brevipinna (Müller & Henle, 1839)	R	R	C	NA
Carcharhinus melanopterus (Quoy & Gaimard, 1824)			VR	NA
Rhizoprionodon acutus (Rüppell, 1837)			VR	NA
Carcharhinus falciformis (Bibron, 1839)			VR	NE
Carcharhinus limbatus (Valenciennes, 1839)			VR	DD
Carcharhinus obscurus (Le Sueur, 1818)			VR	DD
Prionace glauca (Linnaeus, 1758)	C	C	R	CR **
Sphyrnidae				

	GFCM Sub area			IUCN Status
Species	**GSA 12**	**GSA 13**	**GSA 14**	
Sphyrna zygaena (Linnaeus, 1758)		VR	VR	CR
Squaliformes				
Squalidae				
Squalus acanthias Linnaeus				EN
Squalus blainville (Risso, 1827)	VC	VC	C	DD
Squalus megalops (Macleay, 1881)	R	R	C	DD
Somniosidae				
Somnosus rostratus (Risso, 1827)	VR			DD
Dalatiidae				
Dalatias licha (Bonnaterre, 1788)	VR	VR		VU
Etmopteridae				
Etmopterus spinax (Linnaeus, 1758)	R	VR		LC
Oxynotidae				
Oxynotus centrina (Linnaeus, 1758)	C	C	R	CR *
Centrophoridae				
Centrophorus cf. *uyato* (Rafinesque, 1810)	R	R	VR	NE **
Squatiniformes				
Squatinidae				
Squatina aculeata Cuvier, 1829	R	R	R	CR *
Squatina oculata Bonaparte, 1840	R	R	R	CR *
Squatina squatina (Linnaeus, 1758)	R	R	R	CR *
Rajiformes				
Torpedinidae				
Torpedo nobiliana Bonaparte, 1835	R	R		LC
Torpedo marmorata Risso, 1810	C	C	C	LC
Torpedo torpedo (Linnaeus, 1758)	VC	VC	VC	LC
Rhinobatidae				
Glaucostegu cemiculus Geoffroy St. Hilaire, 1817	R	C	VC	EN *
Rhinobatos rhinobatos (Linnaeus, 1758)	R	C	VC	EN *
Rajidae				
Dipturus oxyrinchus (Linnaeus, 1758)	C	C	R	NT
Rostraraja alba Lacépède, 1803	R	R	R	EN *
Leucoraja circularis (Couch, 1838)	VR			CR *
Leucoraja melitensis Clark, 1926	R	R	R	CR *
Raja asterias Delaroche, 1809	VR			NT
Raja clavata Linnaeus, 1758	VC	VC	VC	NT
Raja miraletus Linnaeus, 1758	C	C	C	LC

	GFCM Sub area			IUCN Status
Species	**GSA 12**	**GSA 13**	**GSA 14**	
Raja montagui Fowler, 1910		VR		LC
Raja radula Delaroche, 1809	VC	VC	C	EN
Raja polystigma Regan, 1923	R	R		LC
Raja brachyura Lafont, 1873	VR	VR		NT
Dasyatidae				
Dasyatis centroura (Mitchill, 1815)	R	R	R	VU
Dasyatis marmorata (Steindachner, 1892)	VR	VR	C	DD
Dasyatis pastinaca (Linnaeus, 1758)	C	C	VC	VU
Dasyatis tortonesei Capapé, 1975	C	C	VC	NA
Pteroplatytrygon violacea (Bonaparte, 1832)	C	C	R	LC
Taeniura grabatus (Geoffroy St. Hilaire, 1817	C	C	C	DD
Gymnuridae				
Gymnura altavela (Linnaeus, 1758)	R	R	C	CR *
Myliobatidae				
Myliobatis aquila (Linnaeus, 1758)	R	R	R	VU
Aetobatidae				
Aetomylaeus bovinus (Geoffroy St. Hilaire, 1817	C	C	VC	CR
Mobulidae				
Mobula mobular (Bonnaterre, 1788)	R	C	R	EN *

	Not observed
(red)	Very Rare: Species observed accidentally in the region. Cited no more than three times in the literature.
(yellow)	Rare: Species observed in the region but in a restricted group or in isolated specimens.
(blue)	Common: Species captured in more or less abundant quantities in at least one sector of the region and during a period of the year.
(green)	Very common: Species caught very frequently throughout the region throughout the year
*	Species listed in the annex II of the SPA/BD Protocol to the Barcelona Convention
**	Species listed in the annex III of the SPA/BD Protocol to the Barcelona Convention

Table 1.
Diversity and status of elasmobranchs species occurring in Tunisian water during the last 20 years.

country (GSA 12). Only 10 publications covered the Eastern coast (GSA12) (**Figure 8**). Some studies concern all Tunisian coasts because of the uses of samples from all the countries without distinction between GSA.

Studies concern essentially biology (sexual maturity size, reproductive cycle, size at birth fecundity, etc.), ecology (diet composition, frequency of prey, etc.), and growth (Von Bertalanffy growth parameters, age at maturity, , etc.). Recently, an attention to the impact of fishery, bycatch, and systematic were observed [11–20, 28–30] (**Figure 9**).

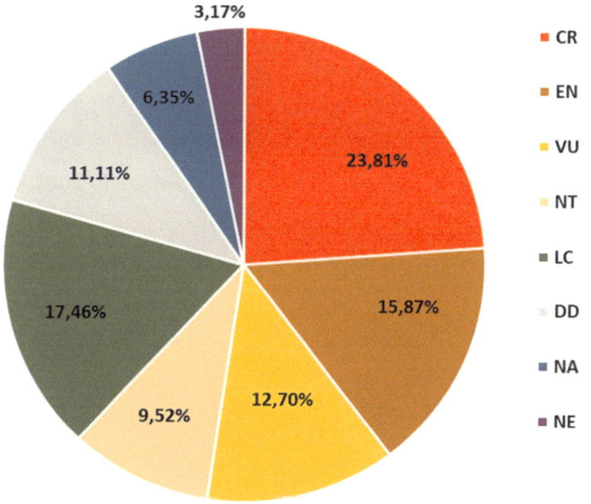

Figure 5.
Species status according to IUCN red list classification.

Figure 6.
Some vulnerable species captured accidentally in Tunisia. 1: Isurus oxyrinchus*; 2:* Gymnura altavela*; 3:* Alpoias vulpinus*; 4:* Aetomylaeus bovinus*; 5:* Raja radula.

5.1 Available data on reproduction

Elasmobranchs are a vulnerable group because of their life histories including the late maturity, the low fecundity, and a long reproductive cycle [31]. Reproductive parameters are crucial to develop conservation strategies and management plan. In Tunisia, data on reproductive parameters are available for 39 species (**Table 2**). However, recent data concern only 16 species. Among them, six species are listed in annex II and III of the of the SPA/BD Protocol to the Barcelona Convention. Reproductive studies related to GSA 13 are scare. The main reproductive parameters are listed in **Table 2**.

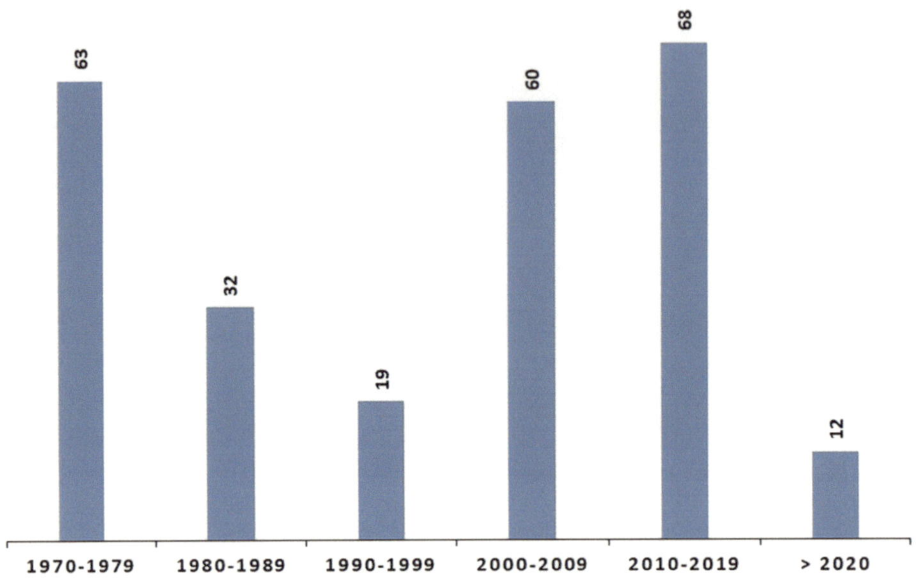

Figure 7.
Temporal distribution of the number of published papers dealing with elasmobranchs in Tunisia.

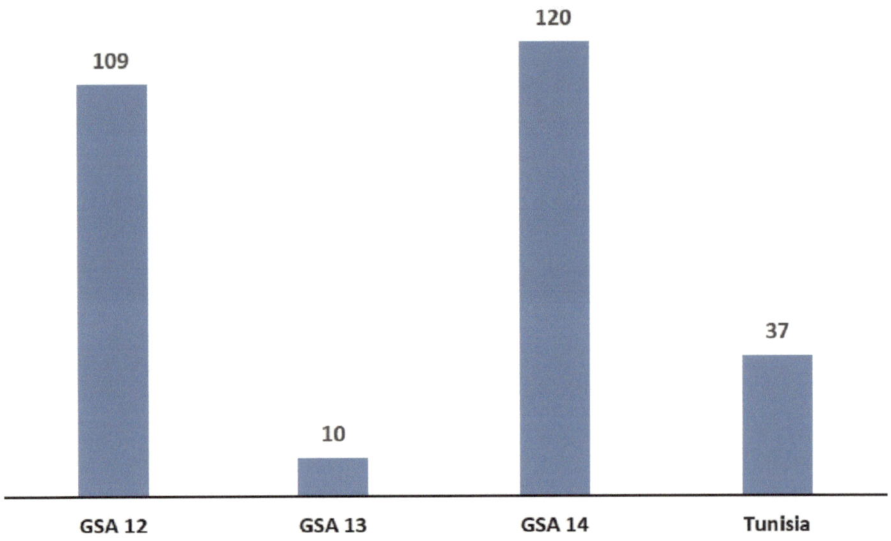

Figure 8.
Geographic distribution of elasmobranchs paper in Tunisia according to GFCM sub-area (1971–2022).

5.2 Available data on age and growth

The age and growth parameters of a population are very important for conservation and management plans [34]. The parameters are used for the determination of natural mortality and longevity and, ultimately for the calculation of vital rates in demographic models [35].

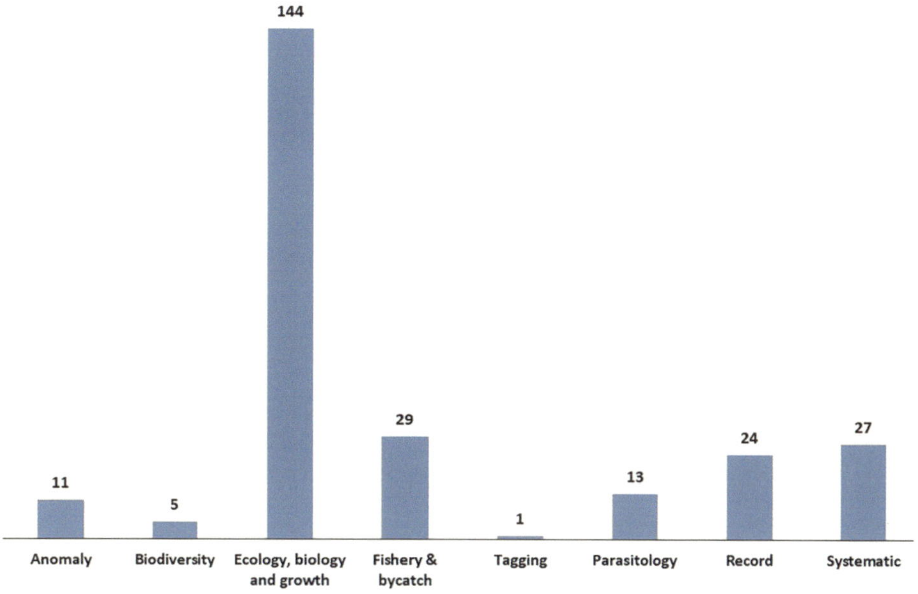

Figure 9.
Distribution of elasmobranchs paper by topic in Tunisia (1971–2022).

For age determination of elasmobranchs in Tunisia, vertebral sections and dorsal spines are used (**Figure 10**). These structures tend to accumulate calcified growth material as they age, thus producing concentric areas that often have characteristics reflecting the time of year in which this material is being deposited [36].

The age and growth studies in Tunisia are recent. They concern the south coast of the country (GSA 14). Age and growth data presented in this section include parameters of the Von Bertalanffy growth model (VBGM) (von Bertalanffy 1938) of eight species: three viviparous species and five oviparous species (**Table 3**).

5.3 Available data on food and feeding habits

Studies of feeding habits are essential to understand the functional role of fish in the ecosystem. Data on feeding can provide information on species distribution and its position in food webs.

Sharks are considered top predators and have an important role in the marine ecosystems. Information about the food habits is essential to appreciate the species biology and ecology, since the quality and quantity of food directly affect species maturation and growth.

In Tunisia, available data on food and feeding habits concern 24 species. Among them 16 species were subject of recent studies mainly in Gabes Gulf (GSA 14). Information on diet composition are summarized in **Table 4**.

5.4 Available data on fishery and by catch

The studies on fisheries and bycatch of elasmobranchs in Tunisia are recent. The first study back to 2003 [113] (**Figure 11**). The low economic value of this group's products compared to bony fishes, crustacean, and mollusk has resulted in a lower priority for research and conservation of these species in Tunisia. It is to highlight that

Scientific name	GSA	Size at maturity (cm)	Gestation (months)	Fecundity	Size at Birth (cm)	References
Heptranchias perlo	12	M:81TL/ F: 85 TL	10	2–18	—	[32]
Mustelus mustelus	12	M:108TL /F: 123 TL	11–12	4–22	34–42	[33]
	14	M: 97 TL/F:118 TL	11–12	4–18	30–40	[8–38]
Mustelus punctulatus	12	M: 87 TL /F: 100 TL	12	5–30	40–43	[39]
	14	M: 81 TL/F: 95 TL	12	12–27	40–43	[8–40]
Mustelus asterias	12	M: 75 TL/F: 96 TL	12	10–35	28–32	[41]
Galeorhinus galeus	TN	M: 126 TL/ F:140 TL	8–36	8–30	24–32	[42]
Carcharhinus brevipinna	TN	M: 172TL/ F: 176TL	13–14	6–10	61–69	[43]
Carcharhinus limbatus	TN	M: 167TL/ F: 178 TL	12	6–8	61–65	[44]
Carcharhinus plumbeus	14	M: 160TL /F: 172 TL	12–14	7	50–65	[7–45]
Oxynotus centrina	12	M: 60 TL/F: 65 TL	12	10–15	21–24	[46]
Squalus blainvillei	12	M: 55TL/F: 60 TL	12	2–6	20–23	[47]
	14	M: 52 TL/F: 62 TL	12	1–6	23–24	[48]
Squalus megalops	14	M: 44 TL/F: 56 TL	12	1–6	18–23	[49]
Etmopterus spinax	12	M: 28 TL/F: 34 TL	—	5–18	9–11	[50]
Squatina aculeata	12	M:120 TL/ F:137 TL	12	8–12	30–35	[51]
Squatina oculata	12	M: 70 TL/F: 90 TL	—	5–8	—	[52]
Squatina squatina	12	M: 80TL/F: 128 TL	—	7–18	—	[52]
Rhinobatos rhinobatos	14	M:75 TL/ F: 887 TL	8	4–12	31	[53, 54]
Glaucostegus cemiculus	14	M:112 TL/F:138 TL	8	4–12	39	[54, 55]
Dasyatis centroura	TN	M: 80/F: 100DW	—	2 to 6	—	[56]
Dasyatis pastinaca	12	M: 32DW/F:38DW	4	3 to 6	12	[57]
	14	M:33 DW/F:40DW	12	10	11–13	[58]
Dasyatis tortonesei	12	M:30 DW/F:32DW	4	3 to 8	—	[59]
	14	M:38DW/F:46DW	12	3 to 10	15.7	[60]
Dasyatis marmorata	14	M:30 DW/F:32DW	3 to 4	2–4	11.8	[61]
Pteroplatytrygon violacea	TN	M: 42DW/F:45DW	4 to 5	2–7	—	[62]
Torpedo nobiliana	12	M: 55 TL/F: 90 TL	12	—	17–22	[63]
Torpedo torpedo	12	M: 19 TL/F: 19 TL	4	1–9	8–9	[64]
	14	M: 23 TL/F: 23 TL	6	7.15	—	[65]
Torpedo marmorata	12	M: 26 TL/F: 40 TL	36	2–17	—	[66]
	14	M:27.5 TL/F: 34 TL	—	8	—	[54]
Aetomylaeus bovinus	12	M:80 DW/F:90DW	8	2–6	—	[67]
Myliobatis aquila	12	M:50 DW/F:70DW	12	8–12	21–29	[68]
Gymnura altavela	12	M:78DW/F:108DW	9	6–8	29	[56]
Scyliorhinus canicula	12	M: 40 TL/F: 35 TL	—	38–190	—	[69]
	13	M: 35 TL/F: 35 TL	—		—	[70]

Scientific name	GSA	Size at maturity (cm)	Gestation (months)	Fecundity	Size at Birth (cm)	References
Scyliorhinus stellaris	12	M: 77 TL/F: 82 TL	—	77–109	—	[71]
Galeus melastomus	12	M: 36TL/F: 49 TL	—	15–29	—	[72]
Raja alba	12	M:91DW/ F:98 DW	—	—	—	[73]
	14	M:94DW/ F: 95DW	—	6–16	—	[74]
Raja asterias	12	M: 51TL/F: 56 TL	—		—	[75]
Raja miraletus	12	M: 54 TL/F: 57 TL	—	—	—	[76]
	14	M: 34TL/F: 41 TL	—	12–60	—	[77]
Raja melitensis	12	M: 40/F: 40	—	10–56	—	[78]
Raja radula	12	M: 68 TL/F: 71 TL	—		—	[79]
	14	M: 47TL/F: 57 TL	—	120	—	[80]
Raja clavata	12	M: 75 TL/F: 85 TL	—	108–262	—	[81]
	14	M: 65 TL/F: 79 TL	—	36–144	—	[82]
Raja polystigma	12	M: 53/F: 63	—		—	[83]

Table 2.
Reproductive parameters of elasmobranch species in Tunisia. TL: Total length, disk width.

some other projects studying bycatch are currently in execution in Tunisia as the "Medbycatch" project.

In Tunisia, elasmobranchs are caught accidentally by all fishing gear (trawl, trammel net, longline, purse senne, etc.) and as a targeted species during the summer by a specific gill nets in the southern coast (GSA 14).

5.4.1 Bycatch of elasmobranchs by longline

In the frame of ACCOBAMS-GFCM Project on mitigating interactions between endangered marine species and longline fishery in Zarzis (GSA 14), developed with the collaboration of the RAC/SPA and a substantial financial support from the MAVA foundation, results show that 46% of the production of the bottom longline targeting groupers are elasmobranchs. Eight sharks and nine batoids are caught by bottom longline. Among elasmobranchs species captured the blackchin guitarfish (*Glaucostegus cemiculus*), the hound sharks (*Mustelus spp.*), the Shortnose spurdog (S. megalops), the sandbar shark (*C. plumbeus*) and the stingray *(Dasyatis spp.)* were the most caught [114, 115] (**Figure 12**).

Elasmobranchs represent more than 90% of the capture of pelagic longline in the southern coast of Tunisia. Nine elasmobranch species were captured by this gear (Sandbar shark, spinner shark, shortfin mako shark, smooth hound, pelagic stingray, blackchin guitarfish, bull rays, round stingray, and thornback ray). The captures were dominated by the sandbar shark accounting about 82.5% of capture [11, 114, 115] (**Figure 13**).

5.4.2 Bycatch of elasmobranchs by trammel nets

Landing monitoring of boat using shrimp's trammel nets in Sfax port during May, June, and July 2009 shows that seven species elasmobranchs (*M. mustelus*, *Mustelus*

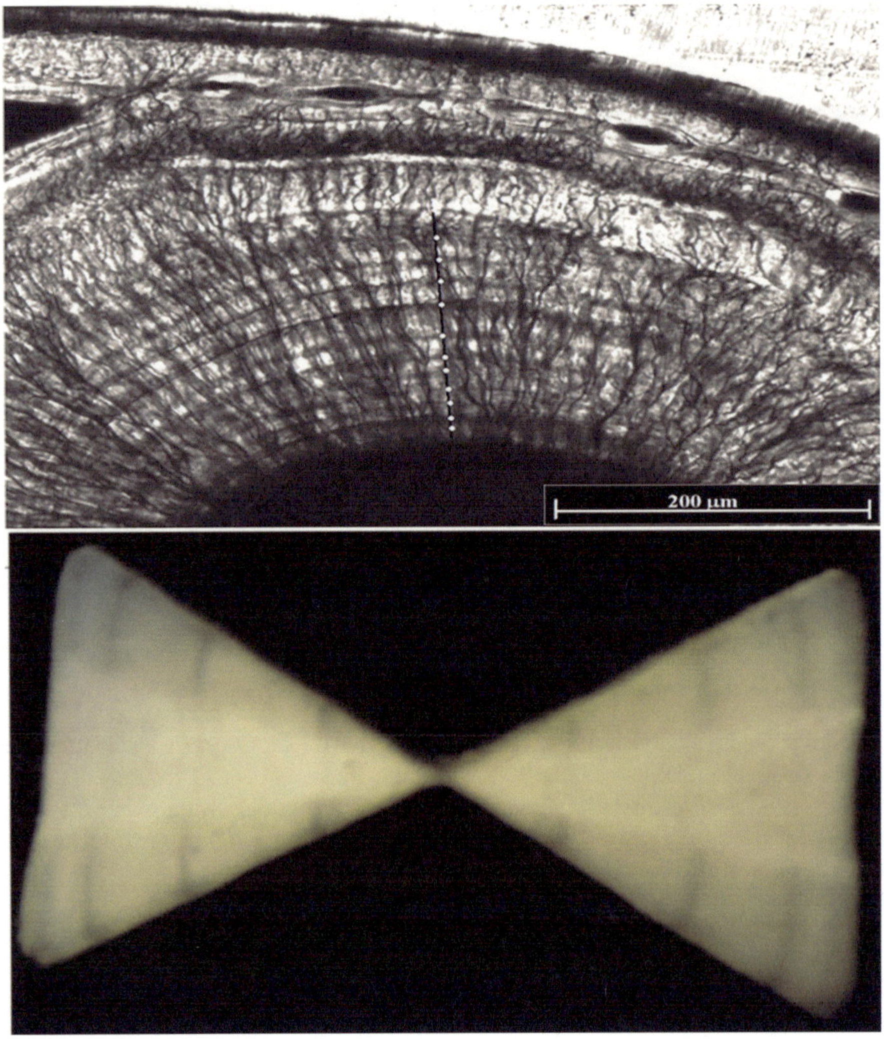

Figure 10.
A thin-section of a longnose spurdog spine and a blackchin guitarfish vertebra from the Gabes Gulf (GSA 14).

punctulatus, *Dasyatis pastinaca*, *Dasyatis marmorata*, *T. torpedo*, *C. plumbeus*, and *Carcharhinus brevipinna*) were caught as by-catch in GSA 14. Elasmobranch by-catch was dominated by sharks (90.3%). The smoothhound sharks *Mustelus spp.* being by far the most important (88.9%) and reflecting their abundance in the area; 58% of the sets caught at least one specimen. Captures were composed essentially of neonate and juvenile sharks, while the batoids were dominated by mature individuals [28].

5.4.3 Bycatch of elasmobranchs by trawl

Monitoring of trawler fishery in the Gulf of Gabes during 2009 shows that Elasmobranchs are commonly caught as by-catch by bottom trawlers in the Gulf of Gabes (GSA 14). A total of 31 elasmobranch species was recorded in trawl captures, among them 14 sharks and 17 batoides representing 64.58% of elasmobranch species observed in the area. Elasmobranch bycatch averaged 5.42% of the total landing

Species	GSAs	Sex	VBGM parameters			t_{max}	A_{mat}	References
			L_∞ (cm)	K	t_0			
Squalus blainvillei	14	M	91.1	0.14	−1.42	15	4.97	[84]
		F	105.7	0.11	−1.12	19	7.44	
Squalus megalops	14	M	68.55	0.08	−4.65	26	8.39	[85]
		F	82.31	0.06	−3.89	29	15.38	
Rostroaja alba	14	M	177.6	0.06	−1.28	32	19.69	[74]
		F	199.6	0.04	−1.47	35	23.47	
Raja radula	14	M	76.35	0.22	−0.16	9	3.39	[80]
		F	97.94	0.14	−0.35	12	5.52	
Glaucostegus cemiculus	14	M	181.6	0.272	−0.71	10	2.89	[55]
		F	200	0.202	−0.81	14	5.09	
Raja clavata	14	M	100.8	0.14	−1.13	12	5.3	[82]
		F	114.6	0.11	−1.23	15	7	
Raja miraletus	14	M	67	0.22	−1.01	7	2.7	[77]
		F	69.2	0.18	−0.11	9	4.41	
Dipturus oxyrinchus	14	M	102.1	0.12	−1.18	22	11.95	[86]
		F	123.9	0.08	−1.26	25	13.96	

Table 3
Von Bertalanffy growth model (VBGM) parameters: L_∞: cm (TL), k: (year-1), t_o (years); t_{max}: oldest fish (years), A_{mat}: age at maturity (years).

(1.7% sharks and 3.7% batoides). The CPUE was estimated at 0.8 Kg/haul for all elasmobranchs. Sharks represented 0.27 Kg/haul and batoides constituted 0.54 Kg/haul. Specimens caught were mainly juveniles [116].

5.4.4 Bycatch of elasmobranchs by purse seine

The purse seine caught a very low proportion elasmobranch especially pelagic sharks and rays. *Mobula mobular*, *I. oxyrinchus*, and *Alopias vulpinus* were the most reported species [117, 118].

5.4.5 Specific fishery

From March to August and between Jerba Island and Zarzis (Southern Tunisia, GSA14) adults of the blackchin guitarfish, the smooth-hound shark, and the sandbar shark are targeted by a little flotilla of small-scale vessel using specific gillnets called locally “Gattatia” and “Garracia”; “Gattatia” for smouth-hound sharks and “Garracia” for Blackchin guitarfish and sandbar shark. Gillnets are in polyamide monofilament netting with a stretched mesh size of 120–160 mm for the first one and 300–400 mm for the second gillnet type. These nets are used at 5–30 m depth. Size composition of captures varied by species, but usually mature, mainly gravid females were abundant [20] (**Figure 14**).

Species	*GSAs*	*Frequency of Prey*					References
		Fish	Cr	Mol	Chon	An	
Heptranchias perlo	12	xxx	xx	x	x		[37]
Squalus blainvillei	14	xxx	xxx	x		*	[87]
Centrophorus granulosus	12	xxx	xx				[88]
Galeus melastomus	12	x	xxx	x	*	*	[89]
Carcharhinus plumbeus	14	xxx	x	xx	x	*	[90]
Mustelus mustelus	14	xx	xxx	x		*	[91]
Mustelus punctulatus	12/14	xx	xxx	x		*	[40–92]
Scyliorhinus canicula	12/13	xx	xxx	x		*	[93, 94]
Scyliorhinus stellaris	12	xx	xxx	x		*	[95]
Rhinobatos rhinobatos	14	xxx	xxx	x		*	[96, 97]
Glaucostegus cemiculus	14	xxx	xxx	x		*	[96, 97]
Dasyatis pastinaca	12/14	xx	xxx	x		x	[60–98]
Dasyatis tortonesei	12/14	xxx	x	x	*		[60–99]
Dasyatis marmorata	14	x	x	xxx		x	[100]
Torpedo torpedo	12/14	xxx	x	—		*	[54–101]
Pteromylaeus bovinus	14	xxx	x	xxx		*	[102]
Myliobatis aquila	12	—	—	xxx		*	[103]
Raja alba	12/14	xxx	x	x	x		[104, 105]
Raja asterias	12	xx	xxx	x		*	[106]
Raja miraletus	12/14	x	xxx	x		*	[107, 108]
Raja melitensis	12		xxx				[109]
Raja radula	12/14	x	xxx	x		*	[107–110]
Raja clavata	12/14	xx	xxx	x	*	*	[111, 112]
Raja polystigma	12	xx	xxx	x		x	[113]

*xxx: Main preys, xx: Secondary preys, x: Accessory preys, *: Accidental preys, fish: Teleost fishes, chon: Chondrichtyens, Mol: Mollusks, an: Annelids, Cr: Crustaceans, other: Other invertebrates.*

Table 4.
Diet composition of elasmobranch species from Tunisia.

6. Critical area

Elasmobranch nurseries are areas characterized by the presence of neonates, small juveniles, and pregnant females. This area offers a better source of food and protection against predation; overall, they are located in coastal, shallow, and highly productive waters. At least four elasmobranch species were perceived use the coastal water of the southern coast of Tunisia (GSA 14) as nursery (**Figure 15**). *C. plumbeus*, *M. mustelus*, *R. rhinobatos*, and *Rhinobatos cemiculus* use the area as a year-round primary and secondary nursery, with juveniles remaining in it up to the size at maturity [7–9].

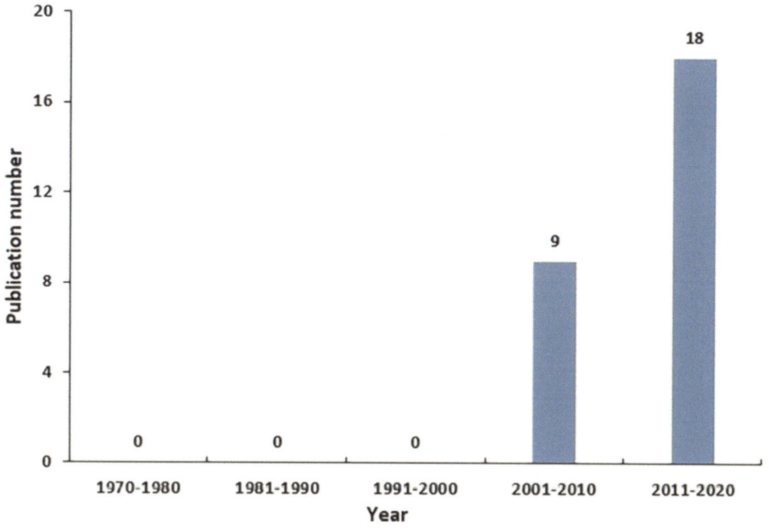

Figure 11.
Temporal distribution of published papers dealing with fishery and bycatch of elasmobranchs in Tunisian coast.

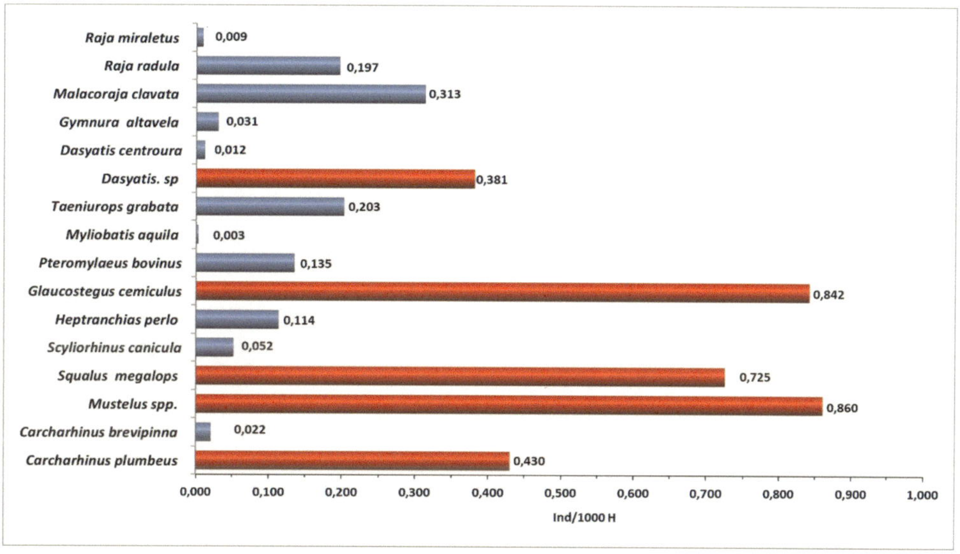

Figure 12.
Elasmobranchs catch rates in the bottom longline in Zarzis zone.

7. Regulations for the protection of elasmobranchs in Tunisia

Tunisia ratified many international conventions dealing with cartilaginous fishes and biodiversity in general (**Table 5**) and adopted the GFCM recommendations on the management and conservation of sharks and rays in the GFCM area of application (Rec. GFCM/36/2012/3; Rec. GFCM/42/2018/2).

The protection of elasmobranchs species is ensured at the national level by the decree n° 94–13 on July 31, 1994 and the decree of September 28, 1995 of the Minister

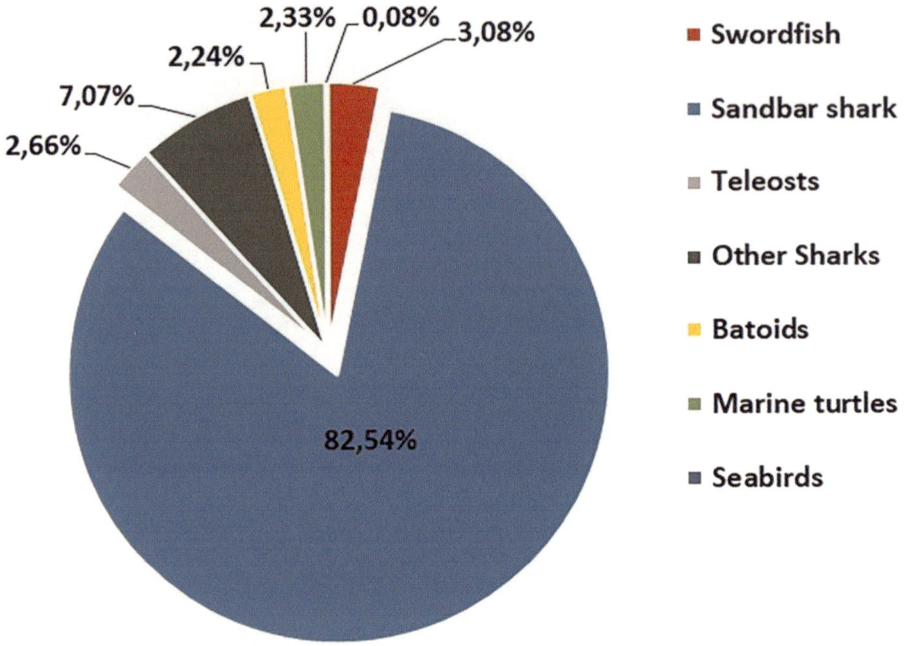

Figure 13.
Catch composition of pelagic longline in Zarzis zone [33].

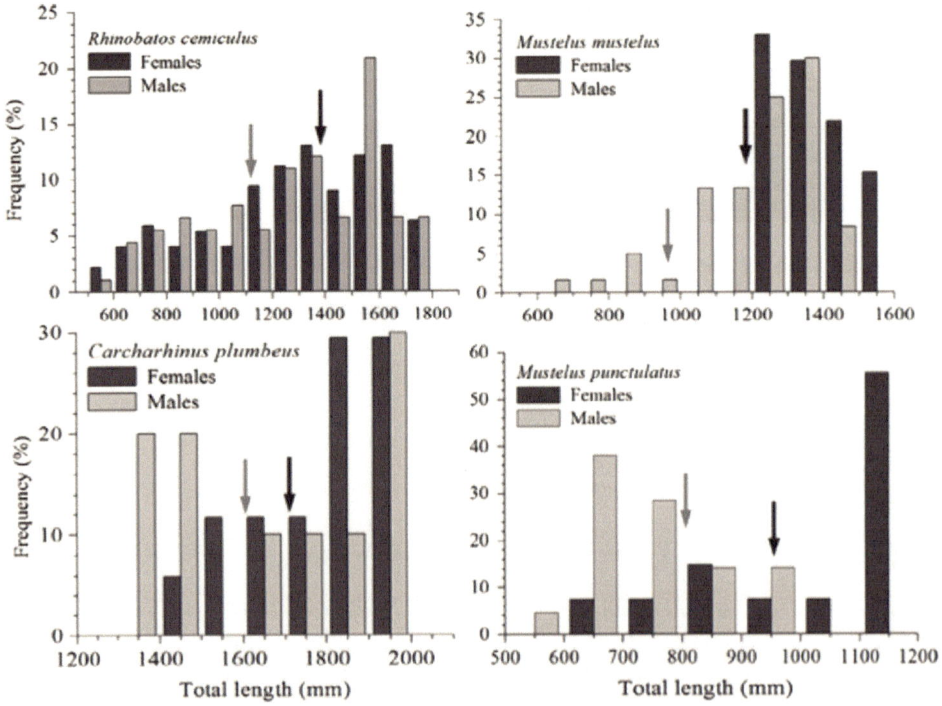

Figure 14.
Length-frequencies distribution of elasmobranch species caught by gillnet [20].

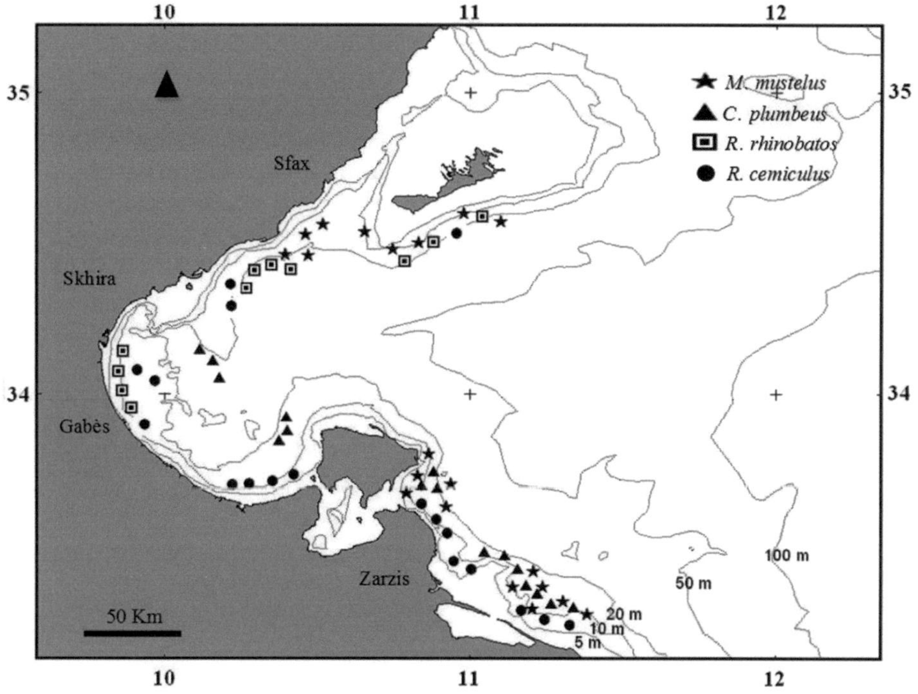

Figure 15.
Nursey area of some elasmobranchs in Gabes Gulf.

Convention	Adoption	Ratification	Loi n.°
CITES	1973	1974	74–12 du 11/05/74
Barcelone	1976	1977	77–29 du 25/05/77
CMS	1979	1986	86–63 du 16/07/86
Berne	1979	1995	95–75 du 07/08/95
CBD	1992	1993	93–45 du 29 /12/ 1994

Table 5.
International conventions ratified by Tunisia.

of Agriculture regulating the practice of fishing activities. The former one is currently being amended to mainly consider the conventions ratified by Tunisia and the recommendations of the CGPM.

8. Recommendations

Sharks and rays occupy a high level in the trophic webs and are characterized by a K-strategy. This determines a high sensibility to fishing pressure. To conserve the biodiversity of this emblematic groups, many actions should be ensured in Tunisia and in the Mediterranean as many species are migratory.

To overcome this situation, it is necessary to improve data collection at sea and at land for a global map of species distribution and for effective landing statistics in all

Tunisian coast; statistic data must be done by species or at least by group of species. For this, it is necessary to focus on species identification trainings and to develop studies on systematic, launched a monitoring to delimit critical area for elasmobranchs in the area and to determine the discard quantity of elasmobranchs by different fishing gear. Experimentation of mitigation measures must be enlarged.

Developing of stock assessment studies; some knowledge on biologic parameters is now available and on fishery; at regional levels because of the urgent need for protection of these groups. Likewise, undertake studies on migration and exchange between populations by satellite tracking.

9. Conclusions

Elasmobranchs represent an average of 2% of the Tunisian national fish production. According to FAO statistics, a mean of 2370 tons per year is landed during the last 20 years. They are landed mainly by small-scale vessels using gillnets, trammel nets, and longlines followed by trawling.

Two hundred and fifty-four references, dealing with elasmobranchs in Tunisia, were analyzed in this chapter.

This analysis shows that 63 elasmobranchs occurred in the area during the last 20 years: 37 sharks belonging to 17 families and 26 batoids belonging to eight families. Three species were observed for the first time in the area during the considered period: *S. megalops, S. rostratus*, and *A. superciliosus.* The southern waters of Tunisia were characterized by the presence of costal species, whereas deep species were found mainly in northern zone. More than 52% of elasmobranchs species observed in Tunisian water were criterial endangered, endangered, or threatened. The Gabes Gulf represents an important area for elasmobranchs, four species use the coastal water of the area as a nursery.

In Tunisia, information on reproduction is available for 40 species. However, recent data concern 16 species. Studies on age and growth concerned only species from the south coast of the country (GSA 14). Von Bertalanffy parameters are available for eight species. Concerning food habits, recent data concern 16 species. Therefore, it is urgent to initiate the study of the age and growth of other species.

Bycatch has become one of the issues to be considered in any development of fisheries. Elasmobranchs which are considered mainly as bycatch are very sensitive given their particular biological characteristics. In Tunisia, trammel nets and trawl in the area cause the capture of juveniles while specific gillnets engender the capture of adults and mainly pregnant females.

Elasmobranchs in Tunisia: Status, Ecology, and Biology
DOI: http://dx.doi.org/10.5772/intechopen.108629

Author details

Samira Enajjar[1]*, Bechir Saidi[2] and Mohamed Nejmeddine Bradai[1]

1 National Institute of Science and Technology of the Sea, R16INSTM02 Marine Biodiversity Laboratory, Carthage University, Sfax, Tunisia

2 Faculty of Sciences and Techniques of Sidi Bouzid, Kairouan University, Sidi Bouzid, Tunisia

*Address all correspondence to: enajjarsamira@yahoo.fr

References

[1] Serena F, Abella AJ, Bargnesi F, Barone M, Colloca F, Ferretti F, et al. Species diversity, taxonomy and distribution of Chondrichthyes in the Mediterranean and Black Sea. The European Zoological Journal. 2020; **47**(1):497-536. DOI: 10.1080/ 24750263.2020.1805518

[2] Dulvy NK, Allen DJ, Ralph GM, Walls RHL. The conservation status of Sharks, Rays and Chimaeras in the Mediterranean Sea [Brochure]. Spain: IUCN; 2016

[3] Ferretti F, Myers RA, Serena F, Lotze HK. Loss of large predatory sharks from the Mediterranean Sea. Conservation Biology. 2008;**22**:952-964

[4] Musick JA, Burgess G, Cailliet G, Camhi M, Fordham S. Management of sharks and their relatives (Elasmobranchii). Fisheries. 2000;**25**:9-13

[5] Bradai MN, Quignard JP, Bouaïn A, Jarbouï O, et al. Ichtyofaune autochtone et exotique des côtes tunisiennes : recensement et biogéographie. Cybium. 2004;**28**:315-328

[6] Bradai MN, Saidi B, Enajjar S. Elasmobranchs of the Mediterranean and Black Sea: Status, ecology and biology. In: Bibliographic Analysis. Studies and Reviews. Rome: FAO; 2012. p. 91

[7] Bradaï MN, Saidi B, Bouaïn A, Guelorget O, Capapé C. The Gulf of Gabès (Southern Tunisia, Central Mediterranean): A nursery area for sandbar shark, *Carcharhinus plumbeus* (Nardo, 1827) (Chondrichthyes: Carcharhinidae). Annales Series Historia Naturalis. 2005;**15**:187-194

[8] Saidi B. Les requins du golfe de Gabès: Diversité et éco-biologie de trois espèces à importance économique *Carcharhinus plumbeus*, *Mustelus mustelus* & *M. punctulatus* [thesis]. Sfax: Sfax University; 2008

[9] Enajjar S, Saidi B, Bradai MN. The Gulf of Gabès (Central Mediterranean Sea): A nursery area for sharks and batoids (Chondrichthyes: Elasmobranchii). Cahiers de Biologie Marine. 2015;**56**:143-150

[10] Bradaï MN, Saidi B, Enajjar S, Bouaïn A. The Gulf of Gabès: A spot for the Mediterranean elasmobranchs. In: Başusta N, Keskin Ç, Serena F, Seret B, editors. The Proceedings of the International Workshop on Mediterranean Cartilaginous Fish with Emphasis on Southern and Eastern Mediterranean, Istanbul, 2005. Turkey: Turkish Marine Research Foundation; 2006. pp. 107-117

[11] Saidi B, Enajjar S, Karaa S, Echwikhi K, Jribi I, Bradai MN. Shark pelagic longline fishery in the Gulf of Gabes: Inter-decadal inspection reveals management needs. Mediterranean Marine Science. 2019;**20**(3):532-541

[12] Bradai MN, Saidi B, Enajjar S. Overview on Mediterranean shark's fisheries: Impact on the biodiversity. In: Türkoğlu M, Önal U, Ismen A, editors. Marine Ecology Biotic and Abiotic Interactions. London, UK: INTECH; 2018. pp. 211-230

[13] Apal. Vers une Gestion Intégrée du Littoral Tunisien. Ministère de l'Environnement et du Développement Durable Agence de Protection et d'Aménagement du Littoral; 2015

[14] Azouz A, Ben OS. Les fonds chalutables de la région Est de la Tunisie (de Kélibia à Mahdia). Bulletin de

l'Institut National Scientifique et Technique d'Océanographie et de pêche de Salammbo. 1975;**4**:49-59

[15] Zaara C. Etude de la stabilité du littoral du Golfe de Hammamet, impact de la dynamique sédimentaire sur la morphologie côtière. Tunis Faculty of Science; 1995

[16] Bradai MN, Ghobel M, Bouain A. Aperçu sur l'activité de pêche dans le gouvernorat de Sfax. Cahier de CERES Série géographique. 1995;**1**:211-236

[17] Bradai MN, Bentivegna F, Jribi I, El Ouaer A, Maatoug K, El Abed A. Monitoring of a loggerhead sea turtle, *Caretta caretta*. In: Second Mediterranean Conference on Marine Turtles. 2009. p. 54

[18] Bradai MN. Diversité du peuplement ichtyque et contribution à la connaissance des Sparidés du golfe de Gabès [Thesis]. Sfax: Sfax University; 2000

[19] Karaa S, Bradai MN, Jribi I, Attia El Hili H, Bouain A. Status of cetaceans in Tunisia through analysis of stranding data from 1937 to 2009. Mammalia. 2012;**76**: 21-29. DOI: 10.1515/MAMM.2011.100

[20] Echwikhi K, Saidi B, Bradai MN, Bouain A. Preliminary data on elasmobranch gillnet fishery in the Gulf of Gabes, Tunisia. Journal of Applied Ichthyology. 2013;**29**:1080-1085. DOI: DOI: 10.1111/jai.12022

[21] DGAP. Yearbook of fisheries statistics in Tunisia. 2020.

[22] Marouani S, Chaâba R, Kadri H, Saidi B, Bouain A, Maltagliati F, et al. Taxonomic research on *Squalus megalops* (Macleay, 1881) and *Squalus blainvillei* (Risso, 1827) (Chondrichthyes: Squalidae) in Tunisian waters (Central Mediterranean Sea). Scientia Marina. 2012;**76**(1):97-109

[23] Capapé C, Rafrafi-Nouira S, Diatta Y, Reynaud C. First record of little sleeper shark, *Somniosus rostratus* (Elasmobranchii: Squaliformes: Somniosidae), from the Tunisian coast, Central Mediterranean Sea. Acta Ichthyologica et Piscatoria. 2020;**50**(4): 475-480. DOI: 10.3750/AIEP/02998

[24] Bouzidi I, Enajjar S, Saidi B, Bradai MN. First record of the bigeye thresher shark *Alopias superciliosus* Lowe. In: (Elasmobranchii: Alopiidae) in Tunisian Water (Central Mediterranean Sea). 1841

[25] Capapé C, Rafrafi-Nouira S, El Kamel-Moutalibi O, Boumaïza M, Reynaud C. First mediterranean records of Spinetail Devil Ray, *Mobula japanica* (Elasmobranchii: Rajiformes: Mobulidae). Acta Ichthyologica et Piscatoria. 2015;**45**(2):211-215. DOI: 10.3750/aip2015.45.2.13

[26] Poortvliet M, Olsen JL, Croll DA, Bernardi G, Newton K, Kollias S, et al. A dated molecular phylogeny of manta and devil rays (Mobulidae) based on mitogenome and nuclear sequences. Molecular Phylogenetics and Evolution. 2015;**83**:72-85

[27] White WT, Corrigan S, Yang L, Henderson AC, Bazinet AL, Swofford DL, et al. Phylogeny of the manta and devilrays (Chondrichthyes: Mobulidae), with an updated taxonomic arrangement for the family. Zoological Journal of the Linnean Society. 2017;**182**(1):50-75. DOI: 10.1093/zoolinnean/zlx018

[28] Saidi B, Enajjar S, Bradai NM. Elasmobranch captures in shrimps trammel net fishery off the Gulf of Gabes (Southern Tunisia, Mediterranean Sea). Journal of Applied Ichthyology. 2016;**32**:421-426. DOI: 10.1111/jai.13061

[29] Saidi B, Karaa S, Enajjar S, Echwikhi K, Jribi I, Bradai MN. Effects

of fishing practice changes on shark pelagic longline captures of the Gulf of Gabes, Tunisia. Aquatic Conservation: Mar Fresh water Ecosystem. 2020;**30**: 53-67. DOI: 10.1002/aqc.3226

[30] Saadaoui A, Saidi B, Elglid A, Séret B, Bradai MN. Taxonomic observations on stingrays of the genus Dasyatis (Chondrichthyes: Dasyatidae) in the Gulf of Gabès (Southeastern Mediterranean Sea). Zootaxa. 2016;**2**: 101-113. DOI: 10.11646/zootaxa.4173.2.1

[31] Stevens JD, Bonfil R, Dulvy NK, Walker PA. The effects of fishing on sharks, rays and chimaeras (chondrichthyans), and the implications for marine ecosystems. ICES Journal of Marine Science. 2000;**57**:476-494

[32] Capapé C. Nouvelle description de *Heptranchias perlo* (Bonnaterre.1788) (Pisces, Pleurotremata, Hexanchidae). Données sur la biologie de la reproduction et le régime alimentaire des spécimens des côtes tunisiennes. Bulletin de l'Office National des Pêches Tunisie. 1980;**4**(2):231-264

[33] Saidi B, Bradaï MN, Bouaïn A. Reproductive biology of the smooth–hound shark *Mustelus mustelus* in the Gulf of Gabès (south–Central Mediterranean Sea). Journal of Fish Biology. 2008;**72**:1343-1354. DOI: 10.1111/j.1095-8649.2008.01801

[34] Cortés E. Life history patterns, demography, and population dynamics. In: Carrier JC, Musick JA, Heithaus MR, editors. Biology of Sharks and their Relatives. CRC Press. 2004. pp. 449-470

[35] Goldman KJ. Age and growth of elasmobranch fishes. In: Musick JA, Bonfil R, editors. Elasmobranch Fisheries Management Techniques. Singapore: Asia Pacific Economic Cooperation; 2004. pp. 97-132

[36] Cailliet GM, Smith WD, Mollet HF, Goldman KJ. Age and growth studies of chondrichthyan fishes: The need for consistency in terminology, verification, validation, and growth function fitting. Environmental Biology of Fishes. 2006; **77**:211-228. DOI: 10.1007/s10641-006-9105-5

[37] Capapé C. Observations sur la sexualité, la reproduction et la fécondité de 8 sélaciens pleurotrèmes vivipares placentaires des côtes tunisiennes. Archives de l'Institut Pasteur de Tunis. 1974;**51**:329-344

[38] Capapé C, Quignard JP. Contribution à la biologie des Triakidae des côtes tunisiennes. I. *Mustelus mediterraneus* Quignard et Capapé, 1972: répartition géographique et bathymétrique, migration et déplacement, reproduction, fécondité. Bulletin de l'Office National des Pêches Tunisie. 1977;**1**:103-122

[39] Saidi B, Bradaï MN, Bouaïn A. Reproductive biology and diet of *Mustelus punctulatus* (Risso, 1826) (Chondrichthyes: Triakidae) from the Gulf of Gabès, Central Mediterranean Sea. Scientia Marina. 2009;**73**(2):249-258. DOI: 10.3989/scimar.2009.73n2249

[40] Capapé C. Nouvelles données sur la biologie de la reproduction de *Mustelus asterias* Cloquet, 1821 (Pisces, Pleurotremata, Triakidae) des côtes tunisiennes. Vie et Milieu. 1983;**33**: 143-152

[41] Capapé C, Mellinger J. Nouvelles données sur la biologie de la reproduction du milandre, *Galeorhinus galeus* (Linné, 1758) (Pisces, Triakidae) des côtes tunisiennes. Cahiers de Biologie Marine. 1988;**29**:135-146

[42] Capapé C, Hemida F, Secka AA, Diatta Y, Guelorget O, Zaouali J. Distribution and reproductive biology of

the spinner shark, *Carcharhinus brevipinna* (Mueller and Henle, 1841) (Chondrichthyes: Carcharhinidae). Israel Journal of Zoology. 2003;**49**: 269-286. DOI: 10.1560/DHHM-A68M-VKQH-CY9F

[43] Capapé C, Seck AA, Diatta Y, Reynaud C, Hemida F, Zaouali J. Reproductive biology of the Blacktip shark, *Carcharhinus limbatus* (Chondrichthyes: Carcharhinidae) off West and North African coasts. Cybium. 2004;**28**(4):275-284

[44] Capapé C. Nouvelles données sur la morphologie et la biologie de la reproduction de *Carcharhinus plumbeus* (Nardo, 1827) (Pisces, Carcharhinidae) des côtes tunisiennes. Investigaciòn Pesquera. 1984;**48**:115-137

[45] Capapé C, Seck AA, Quignard JP. Observations on the reproductive biology of the angular rough shark, *Oxynotus centrina* (Oxynotidae). Cybium. 1999;**23**(3):259-271

[46] Capapé C, Quignard JP. Recherche sur la biologie de *Squalus blainvillei* (Risso, 1826) (Pisces, Squalidae) des côtes tunisiennes: Relation taille–poids du corps, du foie et des gonades. Coefficients de condition. Rapports hépato et gonosomatique. Croissance embryonnaire. Archives de l'Institut Pasteur de Tunis. 1980;**57**(4):385-408

[47] Marouani S, Bradai MN, Bouain A. Taille à la maturité sexuelle de *Squalus blainvillei* (Risso, 1826) du golfe de Gabès (Tunisie). Rapport de la Comité Internationale de la Mer Méditerranée. 2007;**38**:536

[48] Marouani S, Kadri H, Bouain A, Bradai MN. Premières données sur la biologiede la reproduction de l'aiguillat au nez court *Squalus megalops* (Macleay, 1881) du Golfe de Gabès (Méditerranée centrale). In: Proceedings of the LRSE International Symposium on Biodiversity and Coastal Ecosystems (BEL03); 26-28 November 2013; Oran (Algerie); Chouikhi A, Boutiba Z, Moouffok S, Boudderbala M, edition; 2013. p. 566- 572

[49] Capapé C, Bradai MN, Seck AA, Diatta Y, Tomasini JA, Quignard JP. Aspects of the reproductive biology of the velvet belly, *Etmopterus spinax* (Elasmobranchii: Squalidae). Bulletin de l'institut national des sciences et technologies de la mer de Salammbô. 2001;**28**:55-64

[50] Capapé C, Diatta Y, Seck AA, Guélorget O, Ben Souissi J, Zaouali J. Reproduction of the sawback angelshark *Squatina aculeata* (Chondrichthyes: Squatinidae) off Senegal and Tunisia. Cybium. 2005;**29**:147-157

[51] Capapé C, Quignard JP, Mellinger J. Reproduction and development of two angel sharks, *Squatina squatina* and *S. oculata* (Pisces: Squatinidæ), off Tunisian coasts: Semi–delayed vitellogenesis, lack of egg capsule and lecithotrophy. Journal of Fish Biology. 1990;**37**:347-356

[52] Enajjar S, Bradaï MN, Bouaïn A. New data on the reproductive biology of the common guitarfish of the Gulf of Gabès (southern Tunisia, Central Mediterranean). Journal of the Marine Biological Association of the United Kingdom. 2008;**88**:1063-1068. DOI: 10.1017/S0025315408001550

[53] Enajjar S. Diversité des Rajiformes et étude éco–biologique de *Rhinobatos rhinobatos* et *Glaucostegus cemiculus* (Famille des Rhinobatidae) du Golfe de Gabès (Tunisie) [thesis]. Sfax: Sfax University; 2009

[54] Enajjar S, Bradai MN, Bouain A. Age, growth and sexual maturity of the

blackchin guitarfish *Rhinobatos cemiculus* in the Gulf of Gabès (southern Tunisia, Central Mediterranean). Cahiers de Biologie Marine. 2012;**53**:17-23

[55] Capapé C. Première données sur le cycle de la reproduction de *Dasyatis centroura* (Mitchill, 1815) et de Gymnura altavela (Linné, 1758) des côtes tunisiennes. Archives de l'Institut Pasteur de Tunis. 1974;**51**:345-356

[56] Capapé C. Contribution à la biologie des Dasyatidae des côtes tunisiennes. I. *Dasyatis pastinaca* (Linne, 1758). Réparation géographique et bathymétrique, sexualité, reproduction, fécondité. Annali del Museo Civico di Storia naturale di Genova. 1976;**81**:22-32

[57] Saadaoui A, Saidi B, Enajjar S, Bradai MN. Reproductive biology of the common stingray *Dasyatis pastinaca* (Linnaeus, 1758) off the Gulf of Gabès (Central Mediterranean Sea). Cahier de Biologie Marine. 2015;**56**:389-396

[58] Capapé C. Contribution à la biologie des Dasyatidae des côtes tunisiennes. III. *Dasyatis tortonesei* Capapé, 1975: Répartition géographique et bathymétrique, sexualité, reproduction, fécondité. Bulletin de l'institut national des sciences et technologies de la mer de Salammbô. 1978;**5**:97-110

[59] Saadaoui A. Statut Taxonomique, Pêche et Eco-biologie de deux pastenagues: *Dasyatis pastinaca* et *D. cf tortonesei* du golfe de Gabès (Tunisie). Thèse de doctorat en Sciences biologiques. [thesis]. Sfax: Sfax University; 2017. p. 192

[60] Capapé C, Zaouali J. Reproductive biology of the marbled stingray, *Dasyatis marmorata* (Steindachner, 1982) (Pisces: Dasyatidae) in Tunisian waters (Central Mediterranean). Journal of Aquariculture and Aquatic Sciences. 1995;**7**:108-119

[61] Hemida F, Seridji R, Enajjar S, Bradai MN, Collier E, Guélorget O, et al. New observations on the reproductive biology of the pelagic stingray, *Dasyatis violacea* (Bonaparte, 1832) (Chondrichthyes: Dasyatidae) from the Mediterranean Sea. Acta Adriatica. 2003; **44**:193-204

[62] Capapé C. Note préliminaire sur la biologie de *Torpedo (Torpedo) marmorata* Risso, 1810 et de *Topedo (Tetronarce) nobiliana* Bonaparte, 1835 des côtes Tunisiennes. Archives de l'Institut Pasteur de Tunis. 1974;**51**:257-267

[63] Quignard JP, Capapé C. Recherches sur la biologie d'un sélacien du golfe de Tunis, *Torpedo torpedo* Linné, 1758 (Ecologie, sexualité, reproduction). Bulletin de l'institut national des sciences et technologies de la mer de Salammbô. 1974;**3**:99-129

[64] Enajjar S, Bradaï MN, Bouaïn A. La reproduction de la torpille ocellée *Torpedo (Torpedo) torpedo* (Linnaeus, 1758) du Golfe de Gabès. Bulletin de l'institut national des sciences et technologies de la mer de Salammbô. 2002;**7**:40-43

[65] La CC. torpille marbrée, *Torpedo marmorata* Risso, 1810 (Pisces, Rajiformes) des côtes tunisiennes: nouvelles données sur l'écologie et la biologie de la reproduction de l'espèce, avec une comparaison entre les populations méditerranéennes et atlantiques. Annales des Sciences Naturelles, Zoologie, Paris. 1979;**13**(1): 79-97

[66] Capapé C, Quignard JP. Contribution à la systématique et à la biologie de *Pteromylæus bovinus* (Geoffroy Saint-Hilaire, 1817), (Pisces, Myliobatidæ) des côtes tunisiennes. Bulletin du Museum National d'Histoire Naturelle, Paris. 1975;**338**(3):1329-1347

[67] Capapé C, Quignard JP. Dimorphisme sexuel et observations biologiques sur *Myliobatis aquila* (L., 1758). Contribution à l'étude du genre Myliobatis, Cuvier, 1817. Annali del Museo Civico di Storia naturale di Genova. 1974;**50**:1-27

[68] Capapé C, Mnasri-Sioudi N, El Kamel-Moutalibi O, Boumaïza M, Ben Amor MM, Reynaud C. Production, maturity, reproductive cycle and fecundity of small-spotted catshark, *Scyliorhinus canicula* (Chondrichthyes: Scyliorhinidae) from the northern coast of Tunisia (Central Mediterranean). Journal of Ichthyology. 2014;**54**(1):111-126. DOI: 10.1134/S0032945214010020

[69] Saidi B, Bradaï MN, Bouaïn A. Taille à la maturité sexuelle de *Scyliorhinus canicula* des côtes Est et Sud–est de la Tunisie. Rapport de la Comité Internationale de la Mer. Méditerranée. 2004;**37**:433

[70] Capapé C. Contribution à la connaissance de la biologie des Scyliorhinidae des côtes Tunisiennes. III. *Scyliorhinus stellaris* (Linné, 1758). Acta Adriatica. 1977;**17**:1-21

[71] Capapé C, Zaouali J. Contribution à la biologie des Scyliorhinidae des côtes tunisiennes– VI–*Galeus melastomus* Rafinesque, 1810 – Répartition géographique et bathymétrique, sexualité, reproduction, fécondité. Cahiers de Biologie Marine. 1977;**18**:449-463

[72] Capapé C. Contribution à la biologie des Rajidæ des côtes tunisiennes. *Raja alba* Lacépède, 1801: Répartition géographique et bathymétrique, sexualité, reproduction, fécondité. Annexe de l'Institut de Michel Pacha. 1976;**9**:23-47

[73] Kadri H, Marouani S, Bradai MN, Bouain A, Morize E. Age, growth, mortality, longevity and reproductive biology of the White skate, *Rostroraja alba* (Chondrichthyes: Rajidae) of the Gulf of Gabès (Southern Tunisia, Central Mediterranean). Turkish Journal of Fisheries and Aquatic Sciences. 2014;**14**: 193-204. DOI: 10.4194/1303-2712-v14_1_21

[74] Capapé C. Contribution à la biologie des Rajidæ des côtes tunisiennes. IV. Raja asterias Delaroche 1809: répartition géographique et bathymétrique, sexualité, reproduction et fécondité. Bulletin du Muséum National d'Histoire Naturelle de Paris. 1977;**435**:305-326

[75] Capapé C, Quignard JP. Contribution à la biologie des rajidae des côtes tunisiennes. I. *Raja miraletus* Linné, 1758: répartition géographique et bathymétrique, sexualité, reproduction, fécondité. Archives de l'Institut Pasteur de Tunis. 1974;**51**:39-60

[76] Kadri H, Marouani S, Saidi B, Bradai MN, Ghorbel M, Bouain A, et al. Age, growth and reproduction of *Raja miraletus* (Linnaeus, 1758) (Chondrichthyes: Rajidae) of the Gulf of Gabès (Tunisia, Central Mediterranean Sea). Marine Biology Research. 2012;**8**: 309-317. DOI: 10.1080/17451000.2011.619546

[77] Capapé C. Contribution à la biologie des Rajidæ des côtes tunisiennes. VII. Raja melitensis Carck, 1926: sexualité, reproduction, fécondité. Cahiers de Biologie Marine. 1977;**18**:177-190

[78] Capapé C. Contribution à la biologie des Rajidæ des côtes tunisiennes. II. Raja radula Delaroche 1809: Répartition géographique et bathymétrique, sexualité, reproduction et fécondité. Archives de l'Institut Pasteur de Tunis. 1974;**51**:211-228

[79] Kadri H, Marouani S, Bradai MN, Bouaïn A. Age, growth and reproductive biology of the rough skate, *Raja radula*

(Chondrichthyes: Rajidae), off the Gulf of Gabès (southern Tunisia, Central Mediterranean). Marine and Freshwater Research. 2013;**64**(6):540-548. DOI: 10.1071/MF12218

[80] Capapé C. Contribution à la biologie des Rajidæ des côtes tunisiennes. III. *Raja clavata* Linné, 1758. Répartition géographique et bathymétrique, sexualité, reproduction et fécondité. Bulletin du Muséum National d'Histoire Naturelle, Paris, Zoologie. 1976;**393**(3):907-922

[81] Kadri H, Marouani S, Saidi B, Bradai MN, Bouain A, Morize E. Age, growth, sexual maturity and reproduction of the thornback ray, *Raja clavata* (L.), of the Gulf of Gabe's (south-Central Mediterranean Sea). Marine Biology Research. 2014;**10**(4):416-425. DOI: 10.1080/17451000.2013.797584

[82] Capapé C, Quignard JP. Contribution à la biologie des rajidae des côtes tunisiennes. 14. *Raja polystigma* Regan, 1923: répartition géographique et bathymétrique, sexualité, reproduction, fécondité. Cahiers de Biologie Marine. 1978;**19**:233-244

[83] Marouani S, Kadri H, Bradaï MN, Saidi B, Bouaïn A. Age, growth and age at sexual maturity of the longnose spurdog, *Squalus blainvillei*, in the Gulf of Gabès (Tunisia). Rapport de la Commission Internationale de la Mer Méditerranée. 2010;**39**:581

[84] Marouani S, Kadri H, Bradai MN. First data on the age, growth and sexual maturity of the piked spurdog, *Squalus megalops* (Chondrichthyes: Squalidae), in the Gulf of Gabe's (Central Mediterranean Sea). Marine and Freshwater Research. 2015;**67**(5):578-585. DOI: 10.1071/MF14356

[85] Kadri H, Marouani S, Bradai MN, Bouain A, Morize E. Age, growth, longevity, mortality and reproductive biology of *Dipturus oxyrinchus*, (chondrichthyes: rajidae) off the Gulf of Gabès (southern Tunisia, Central Mediterranean). Journal of the Marine Biological Association of the United Kingdom. 2015;**95**(3):569-577. DOI: 10.1017/S0025315414000551

[86] Marouani S, Bradaï MN, Saidi B, Bouain A. Contribution à l'étude du régime alimentaire du *Squalus blainvillei* dans le golfe de Gabès. Bulletin de l'Institut des Sciences et Technologies de la Mer de Salammbô. 2007;**11**:78-80

[87] Capapé C. Nouvelle description de *Centophorus granulosus* (Schneider, 1801) (Pisces, Squalidae). Données sur la biologie de la reproduction et le régime alimentaire des spécimens des côtes tunisiennes. Bulletin de l'Institut des Sciences et Technologies de la Mer de Salammbô. 1985;**12**:97-141

[88] Capapé C, Zaouali J. Contribution à la biologie des Scyliorhinidaedes côtes tunisiennes V. *Galeus melastomus* Rafinesque, 1810: Régime alimentaire. Archives de l'Institut Pasteur de Tunis. 1976;**53**:281-291

[89] Saidi B, Bradaï MN, Bouaïn A, Capapé C. Feeding habits of the sandbar shark *Carcharhinus plumbeus* (Chondrichthyes: Carcharhinidae) from the Gulf of Gabès. Tunisia. Cahiers de Biologie Marine. 2007;**48**:139-144

[90] Saidi B, Enajjar S, Bradai MN, Bouain A. Diet composition of smooth–hound shark, *Mustelus mustelus* (Linnaeus, 1758), in the Gulf of Gabès, southern Tunisia. Journal of Applied Ichthyology. 2009;**25**:113-118. DOI: 10.1111/j.1439-0426.2009.01306.x

[91] Capapé C, Quignard JP. Contribution à la biologie des Triakidae des côtes tunisiennes II. *Mustelus*

mediterraneus Quignard et Capapé, 1972: Régime alimentaire. Bulletin de l'Institut des Sciences et Technologies de la Mer de Salammbô. 1977;**1**:173-179

[92] Capapé C. Contribution à la biologie des Scyliorhinidae des côtes tunisiennes II. *Scyliorhinus canicula* Linné, 1758: Régime alimentaire. Annales, Institut Michel Pacha. 1974;**7**:13-26

[93] Saidi B, Bradaï MN, Bouaïn A. Régime alimentaire de la petite roussette *Scyliorhinus canicula* des côtes est et sud–est de la Tunisie. Bulletin de l'Institut National des Sciences et Technologies de la Mer. 2002;7:17-20

[94] Capapé C. Contribution à la biologie des Scyliorhinidae des côtes tunisiennes. IV. *Scyliorhinus stellaris* (Linné, 1758). Régime alimentaire. Archives de l'Institut Pasteur de Tunis. 1975;**52**:383-394

[95] Capapé C, Zaouali J. Etude du régime alimentaire de deux Sélaciens commune dans le golfe de Gabès (Tunisie) : *Rhinobatos rhinobatos* (linnaeus, 1758) et *Rhinobatos cemiculus* Geoffory Saint–Hilaire, 1817. Archives de l'Institut Pasteur de Tunis. 1979;**56**: 285-306

[96] Enajjar S, Bradaï MN, Bouaïn A. Feeding habits of *Rhinobatos rhinobatos* in the Gulf of Gabès. Rapport de la Commité International de la Mer Méditerrannée. 2007;**38**:468

[97] Capapé C. Contribution à la biologie des Dasyatidae des côtes Tunisiennes. II. *Dasyatis pastinaca* (Linne, 1758). Régime alimentaire. Institut Michel Pacha. 1975; **8**:1-5

[98] Capapé C. Contribution à la biologie des Dasyatidae des côtes tunisiennes. IV. *Dasyatis tortonesei* Capapé, 1975: Régime alimentaire. Archives de l'Institut Pasteur de Tunis. 1978;**55**:359-369

[99] Capapé C, Zaouali J. Le régime alimentaire de la pastenague marbrée, *Dasyatis marmorata* (Pisces, Dasyatidae), des eaux Tunisiennes. Vie et Millieu. 1992;**42**:269-276

[100] El Kamel-Moutalibi O, Mnasri N, Boumaiza M, Reynaud C, Capapé C. Diet of common *torpedo Torpedo torpedo* (Chondrichthyes: Torpedinidae) from the Lagoon of Bizerte (northeastern Tunisia, Central Mediterranean). Cahiers de Biologie Marine. 2013;**54**(2): 209-220

[101] Capapé C. Etude du régime alimentaire de la Mourine vachette, *Pteromylaeus bovinus* (Geoffroy Saint–Hilaire, 1817) (Pisces, Myliobatidae) des côtes tunisiennes. Journal du Conseil International pour l'Exploration de la Mer. 1977;**37**(3):214-220

[102] Capapé C. Étude du régime alimentaire de l'aigle de mer, *Myliobatis aquila* (L., 1758) des côtes tunisiennes. Journal du Conseil International pour l'Exploration de la Mer. 1976;**37**(1):29-35

[103] Capapé C. Contribution à la biologie des Rajidae des côtes tunisiennes. XII. *Raja alba* Lacepede, 1803: Regime alimentaire. Archives de l'Institut Pasteur de Tunis. 1977;**54**(54): 85-95

[104] Kadri H, Marouani S, Bouain A, Bradai MN. Régime alimentaire de *Rostroraja alba*, raie blanche, dans le Golfe de Gabès. Rapport de la Comité Internationale de la Mer Méditerranée. 2013;**40**:468

[105] Capapé C, Quignard JP. Contribution à la biologie des Rajidae des côtes tunisiennes. 6 – *Raja asterias* Delaroche, 1809. Régime alimentaire. Bulletin de l'Institut National des Sciences et Technologies de la Mer. 1977; **4**(2–4):319-332

[106] Capapé C, Azouz A. Etude du régime alimentaire de deux raies communes dans le golfe de Tunis: *Raja miraletus* Linné, 1758 et *R. radula*, Delaroche, 1809. Archives de l'Institut Pasteur de Tunis. 1975;**52**:233-250

[107] Kadri H, Marouani S, Bradai MN, Bouain A. Food habits of the brown ray *Raja miraletus* (Chondrichthyes: Rajidae) from the Gulf of Gabe's (Tunisia). Marine Biology Research. 2014;**10**(4): 426-434. DOI: 10.1080/17451000. 2013.797583

[108] Capapé C. Contribution à la biologie des Rajidae des côtes tunisiennes. VIII. *Raja melitensis* Clark, 1926. Régime alimentaire. Archives de l'Institut Pasteur de Tunis. 1975;**1**(2):40-46

[109] Kadri H, Saidi B, Marouani S, Bradai MN, Bouaïn A. Food habits of the rough ray *Raja radula* (Chondrichthyes: Rajidae) from the Gulf of Gabès (Central Mediterranean Sea). Italian Journal of Zoology. 2013;**80**:52-59. DOI: 10.1080/ 11250003.2012.697925

[110] Capapé C. Contribution à la biologie des Rajidae des côtes tunisiennes. IV. *Raja clavata* (Linné 1758). Régime alimentaire. Annexe. Institut Michel Pacha. 1975;**8**:16-32

[111] Kadri H, Marouani S, Saidi B, Bradai MN, Bouain A. Diet and feeding strategy of thornback ray, *Raja clavata* (Chondrichthyes: Rajidae) from the Gulf of Gabès (Tunisia-Central Mediterranean Sea). Journal of the Marine Biological Association of the United Kingdom. 2014;**94**(7):1509-1516. DOI: 10.1017/S0025315414000587

[112] Capapé C. Contribution à la biologie des Rajidae des côtes Tunisiennes-*Raja polystigma* Regan, 1926. Regime alimentaire. Biljeske–Notes, Split. 1979; **35**:1-10

[113] Saidi B, Bradaï MN, Bouaïn A. La pêche des requins dans la région du golfe de Gabès. Bulletin de l'Institut National des Sciences et Technologies de la Mer. 2003;**8**:84-87

[114] Bradai MN, Saïdi B, Enajjar S, Karaa S. Pêcheries aux palangres de fond et de surface dans le Golfe de Gabès: Rapport final. MoU ACCOBAMS N°07/ 2016/LB6410. 2016.

[115] Enajjar S, Saidi B, Karaa S, Chouikhi K, Bradai MN. Vulnerability of elasmobranchs as longline bycatch in the Gulf of Gabès (Southern Tunisia, central Mediterranean Sea). Deuxième Conférence Méditerranéenne «BIODIV » 28 au 30 octobre 2017 (Monastir; TUNISIE). Bulletin de l'Institut National des Sciences et Technologies de la Mer. 2018;**45**:61-64

[116] Hamdaoui B. Les élasmobranches dans les débarquements des chalutiers au port de pêche de Sfax, golfe de Gabès. [Master]. Sfax: Sfax University; 2009

[117] Bouzidi I. Les élasmobranches (Requins & Raies) dans les pêcheries de Kélibia [Master]. Sfax: Sfax University; 2020

[118] Zammit Chatti M. Les élasmobranches, les tortues marines et les cétacés dans les pêcheries de Teboulba [Master]. Sfax: Sfax University; 2020

Chapter 2

Overview of the Genus *Squalus* in the Mediterranean Sea

Sondes Marouani, Sami Karaa and Othman Jarboui

Abstract

In the Mediterranean Sea, in addition to the two historically known species belonging to the *Squalus* genus (*Squalus blainville* and *Squalus acanthias*), a third species, *Squalus megalops*, has been reported. This last specie is a subject of debate between authors. *S. acanthias* is quite distinct from the other species of the genus *Squalus*, while S. *blainville* and *S. megalops* are very similar morphologically. This similarity has resulted in considerable confusion over their taxonomy. The lack of a well-preserved holotype for *S. blainville*, misidentifications in databases and in the literature, description, and figure of Risso (1827) not conforming to any known species of *Squalus* are impediments to the proper taxonomic identification and the potential revision of the genus. This chapter aims to clarify the state of the species of the genus *Squalus* in the Mediterranean Sea, taking into account all the studies carried out on this subject.

Keywords: sharks' misidentification, *squalus* genus, *Squalus blainville*, *Squalus acanthias*, *Squalus megalops*, Mediterranean Sea

1. Introduction

The Mediterranean Sea is a semi-enclosed sea covering less than 1% of the surface of the global oceans, even though it constitutes a general richness hotspot of total species on a global scale [1, 2]. This richness of cartilaginous and bony fishes is likely the result of the recolonization of the Mediterranean basin after the Messinian crisis. As demonstrated for the great white shark [3], pulses of species immigrations occurred during the glacial and interglacial periods of the quaternary [4].

It is a heterogeneous biogeographic area that shows a high level of biological diversity. This sea constitutes a complex marine ecosystem within which elasmobranchs play a basic role in controlling trophic relationships [5]. This is related to multiple factors from its geological history to its peculiar oceanographic and ecological features [6].

The Mediterranean Sea is considered a Chondrichthyes-rich basin. Recently, a total of 88 chondrichthyan species were listed, representing 30 families and 48 genera in the area [6]. This list includes 48 species of sharks, belonging to 18 families and 27 genera, 38 species of batoids, belonging to 11 families and 19 genera, and two chimeras belonging to two different genera.

Despite its richness, it encloses the highest proportion of threatened species in the world [7], in the Mediterranean Sea, where at least 53% of the species are classified by the IUCN as vulnerable, endangered, and critically endangered [8, 9]. Quite a large proportion of species (13%) are still classified as data deficient [8].

There are information gaps with respect to species richness and abundance of elasmobranchs in the Mediterranean Sea. These gaps often make it hard for international organizations to assess the conservation status of populations. Either the knowledge of the abundance and richness of this group, which has played a crucial ecological part in Mediterranean trophic webs, is also significant to any future strategic plan for the conservation of marine biodiversity in the region [10, 11].

The gaps are due to many reasons, the most important of which is the misidentification of species in databases and in the literature. Generally, elasmobranchs have suffered major taxonomic constraints that have led to misidentification issues related to by-catch and fisheries, which were usually solved by grouping data at higher taxonomic levels, such as genus or family [12, 13], or to morphological and biological similarities among some species, which have led to considerable confusion over their taxonomy such the case of the *Squalus* genus species [14].

Dogfish are scientifically classified as the Squalidae family, categorized under the squaliform order, which encompasses seven families in total, including Squalidae. The latter, more commonly known as dog sharks or spiny dogfish, have two dorsal fins different in shape with long spines without grooves and anal fin, with a cylindrical body and a small mouth. Their jaws are furnished with powerful cutting teeth and sometimes present only on the lower jaw; the upper jaw plays, in this case, only the role of holding the prey. The species of this family are generally small sharks, which frequently generally more or less accentuated bottoms except for the spiny dog *S. acanthias,* which does not descend below 150 meters.

Some species are highly valued and important as a major fish resource for food and liver oil. Some species are threatened due to overfishing and because of their biological characteristics, namely a long lifespan, late maturity, and low fecundity, as is the case with all elasmobranchs.

The Squalidae family itself contains two separate genera: *Cirrhigaleus* and *Squalus*, numbering together 37 species, and has the most species in the group, including *Squaliolus laticaudus*, one of the smallest known sharks with a size of 15 cm.

The genus *Squalus* Linnaeus, 1758 is distributed worldwide [15]. Until 2013, 25 species were known: 14 species documented as valid and 11 species added latterly from the western Indo-Pacific ocean [16–18]. But, this number has recently increased due to the resurrection of *Squalus acutipinnis* (Regan 1908) from South Africa and the description of four new species (*S. albicaudatus*, *S. bahiensis*, *S. lobularis,* and *S. quasimodo*) from the southwest Atlantic [19–21].

It was stated that the species diversity within the group is still poorly characterized [22]. For instance, 20 species have been described or resurrected in the last decade in the Indo-Pacific region [23–25] and the south Atlantic [19, 20].

In fact, cryptic speciation among elasmobranchs is very common [26, 27] and the number of new descriptions, redescriptions, and resurrections of species is growing with the increasing application of molecular tools and integrated taxonomic methodologies. Thus, the number of valid species in the genus was doubled and a significant amount of "hidden" diversity in the group has been revealed. Consequently, about 50% of the *Squalus* species are considered data deficient consistent with the International Union for the Conservation of Nature (IUCN) red list of threatened species [22].

The *Squalus* species inhabit the waters of the continental shelf and upper slope, between 300 and 700 m of depth [28–30], as well as some seamounts and the waters around oceanic islands [22, 31]. They have been divided into four assemblages based on their morphology: the "*acanthias*" group, the "*blainville*" group, the "*megalops cubensis*" group, and the "*asper-barbifer*" group [32–34]. However, members of the "*asper-barbifer*" group are assigned to the genus *Cirrhigaleus* [17]. Nevertheless, in recent years, different genetic studies have attempted to identify *Squalus* species using mitochondrial COI and NADH2 genes [22, 35–37]. Generally speaking, three well-defined groups within the genus have been described: group I, including *Squalus* suckleyi and *S. acanthias*; group II, including *S. blainville/S. megalops/Squalus raoulensis/Squalus brevirostris*; and a third group, "the *Squalus mitsukurii* complex" including *Squalus edmundsi*, *Squalus japonicus*, *Squalus grahami*, *Squalus clarkae*, *and S. mitsukurii* [22, 38–40].

In the Mediterranean, in addition to the two historically known species the longnose spurdog *S. blainville* (Risso, 1827) and the spiny dogfish *S. acanthias* (Linnaeus, 1758), a third species, the shortnose spurdog *S. megalops* (Macleay, 1881), has been reported [40–43].

In the following, we try to clarify the state of the species of the genus *Squalus* in the Mediterranean Sea taking into account all the studies carried out on this subject.

2. Status of *squalus* genus in the Mediterranean Sea

2.1 Ecobiology

As many elasmobranchs, Squalidae are K-selected species with slow growth rates, low fecundity, and late sexual maturation; those species tend to aggregate by sex and size [31, 44, 45]. These features make such a taxon greatly vulnerable to overfishing. Thus, increased understanding of their ecobiology is crucial to develop an assessment for conservation strategies and developing effective fisheries management.

	Maximum TL (cm)	Maturity size (cm TL)	Uterine Fecundity	Size at birth (cm TL)	Location, reference
Squalus blainville	♀89.0	51–65	1–5	21–24	Gulf of Tunis, Tunisia [46]
	♂76.0	45–51			
	♀ 84.0	70	2–6	—	Gulf of Tunis, Tunisia [47]
	♂ 64.0	51			
	♀ 75.0	60	—	20–23.5	Gulf of Tunis, Tunisia [48]
	♂ 62.0	55			
	♀ 92.0	57–58	—	15.5–16.5	Strait of Sicily, Italy [29]
	♂ 73.5	45–46			
	♀78.5	60.1	2–6	19–22	Ionian Sea, Greece [49]
	♂66.4	45–51			
	♀100	62.5	2–6	23.2–24.5	Gulf of Gabès, Tunisia [50, 51]
	♂83.4	52.3			
	♀77.9	65.44	1–6	18–20.9	Eastern Mediterranean Sea [52]
	♂79.9	45.77			
	♀ 95.0	60.3	—	—	Eastern Ionian Sea [53]
	♂ 100.0	41.3			

	Maximum TL (cm)	Maturity size (cm TL)	Uterine Fecundity	Size at birth (cm TL)	Location, reference
S. megalops	♀76	56.41	2–6	18.8–23.6	Gulf of Gabès, Central Mediterranean Sea [50, 51]
	♂69	44.39			
S. acanthias	♀111.0	86–88	4–12	—	Languedocian coast, Northern Mediterranean [54]
	♂80.0	63.5–70			
	♀82.0	51.8	1–6	7.2–22	Eastern Mediterranean Sea [55]
	♂75.5	47.0			
	♀117.5	56.4	1–9	22.3	North Aegean Sea [56]
	♂121.6	52.8			
	♀102.5	57.5	1–20	21–22	Adratic Sea [57]
	♂87.5	65.9			

Table 1.
Studies on Squalus blainville, S. megalops, *and* Squalus acanthias *reproduction carried out in the Mediterranean Sea.*

In the Mediterranean Sea, many investigations on the life history traits of Squalidae were conducted toward filling the information gap and to develop practical conservation strategies for those species in the area. As shown in **Tables 1–4**, a survey of the available published literature was carried out through a bibliographic study.

Until now, the historical traits of the Mediterranean population of *S. megalops* remained poorly studied, which may be attributed to its taxonomic problem in the area.

Generally speaking, Squalidae species are aplacental viviparous, with a long gestation, estimated up to 2 years for *S. acanthias* [81]. Their uterine fecundity was estimated to be from 1 to 12 embryos per litter. Females reach maturity at up to 70 cm,

Species	Prey frequencies							Location, reference
	Chond.	Teleo.	Crus.	Moll.	Ann.	Echino.	Oth.	
Squalus blainville		+++	+++	+	+			Gulf of Gabès, Tunisia [51]
		+++	++	+++		+		Eastern Ionian Sea [53]
		+++	+	+++	+		—	Aegean sea [58]
		++	+	+++				Maltese Islands [59]
		+++	+	++	—			Eastern Mediterranean [60]
S. acanthias		+++		+++		+	+	Eastern Ionian Sea [53]
		++	—	+++				western Mediterranean Sea [61]
		+++	+	++				Adriatic Sea [57]
		+++						Sea of Marmara [62]
Squalus megalops	—	+++	+	+	+	+		Gulf of Gabès, Tunisia [63]

Table 2.
Studies on Squalus blainville, S. megalops, *and* Squalus acanthias *diet carried out in the Mediterranean Sea (−; +; ++; +++: Increasing gradient of prey importance according to the calculated IRI%/ Chond.: Chondrichthyes; Teleo.: Teleosteans, crus.: Crustaceans; moll.: Mollusks; Ann.: Annelids; Echino.: Echinoderms; Oth.: Others).*

56.41 cm, and 88 cm total length (TL), whereas males matured at up to 55.0 cm, 44.39 cm, and 70.0 cm TL for *S. blainville*, *S. megalops,* and *S. acanthias,* respectively (**Table 1**). The estimated size at maturity differed between males and females confirming the marked sexual size dimorphism of those species.

Concerning the food habits of those species, they are active predators, which feed on similar preys but with different importance according to the index of relevant importance of each prey item consumed (IRI%) calculated (**Table 2**). Species are reported to feed mainly on bony fishes, cephalopods, and crustaceans.

Sexual size and mass dimorphism were observed, with females attaining larger TL and greater mass than males (**Table 3**). This pattern is common among viviparous

	Sex	N	LWRs	GT	Location, reference
Squalus blainville	M	1038	W = 0.0033 L 3.09	+	Strait of Sicily, Italy [29]
	F	812	W - 0.0037 L 3.07	+	
	C	27	W = 0.0012 L 3.37	+	Balearic Islands [64]
	C	88	W = 0.0035 L 3.06	+	Eastern Adriatic Sea [65]
	M	108	W = 0.002 L 3.37	+	Gulf of Gabès, Tunisia [66]
	F	124	W = 0.003 L 3.10	+	
	C	299	W = 0.00345 L 3.06	+	North Aegean Sea [67]
	C	27	W = 0.0030 L 3.07	I	Saros Bay North Aegean Sea [68]
	M	8	W = 0.0145 L2.68	—	
	F	19	W = 0.0016 L 3.21	+	
	C	18	W = 0.00004 L 2.48	—	Sea of Marmara [69]
	C	177	W = 0.0033 L 3.06	+	Antalya Bay [70]
	M	92	W = 0.0035 L 3.03	I	eastern Mediterranean Sea [52]
	F	85	W = 0.0029 L 3.10	+	
	C	308	W = 0.0048 L 2.96	I	Central Eagean Sea [71]
	M	149	W = 0.0049 L 2.95	I	
	F	159	W = 0.0046 L 2.98	I	
	M	361	W = 2E-06 L 3.16	+	Eastern Mediterranean Sea [72]
	F	445	W = 7E-07 L 3.32	+	
	C	11	W = 0.0819 L 2.89	—	Lebanease marine waters [73]
	C	14	W = 0.0053 L 2.95	I	North Eagean sea [74]
	C	184	W = 0.0010 L 3.35	+	the Levantine Sea, eastern Mediterranean Sea [75]
	M	85	W = 0.0009 L 3.40	+	
	F	99	W = 0.0014 L 3.29	+	
	F	1282	W = 0.003 L 3.08	+	South of Sicily Central Mediterranean Sea [76]
	M	970	W = 0.005 L 2.98	—	
S. megalops	C	630	W = 0.002 L 3.13	+	Gulf of Gabès, Tunisia [77]
	M	323	W = 0.005 L 2.98	—	
	F	307	W = 0.005 L 2.98	—	

	Sex	N	LWRs	GT	Location, reference
S. acanthias	C	32	W = 0.0112 L 2.77	+	North Aegean Sea [78]
	F	16	W = 0.0023 L 3.18	+	
	M	16	W = 0.0014 L 3.29	+	
	C	32	W = 0.0031 L 3.11	+	Aegean Sea [79]
	C	421	W = 0.00201 L 3.15	+	Eastern Adriatic Sea [66]
	F	346	W = 0.0075 L 2.86	—	North Aegean Sea [56]
	M	274	W = 0.003 L 3.11	+	
	C	565	W = 0.0037 L 3.04	+	Saros Bay [68]
	M	253	W = 0.0072 L 2.86	—	
	F	312	W = 0.0027 L 3.12	+	
	C	8	W = 0.00003 L 2.61	—	Sea of Marmara [69]
	M	150	W = 9.54 10–7 L 3.20	+	Adriatic Sea [57]
	F	176	W = 9.5 10–7 L 3.21	+	

Table 3.
Studies on squalus blanville, S. megalops, and Squalus acanthias mass-lenght relationship, carried out in the Mediterranean Sea (N: Number of individuals; M: Male; F: Female; L: Lenght; W: Weight; GT: Growth type; I: Isometry; (+): Positive allometry; (−): Negative allometry).

sharks since for females, due to their more energetically demanding reproductive mode, there is a strong selection pressure for larger body size [82].

Squalidae are long-lived animals, with females attaining greater age than males, as it is typical of most elasmobranchs. Using the von Bertallonfy model, the investigations were conducted to confirm this pattern of differential growth between males and females (**Table 4**). The maximum ages observed for males were 22 years, 26 years, and 23 years and for females were 28 years, 29 years, and 36 years of *S. blainville*, *S. megalops,* and *S. acanthias,* respectively.

2.2 Geographic distribution

S. acanthias and *S. blainville* are mostly found in the northern part of the Mediterranean Sea and the Adriatic, including the Black Sea [30, 83–85]. Some authors have reported the spiny dogfish (*S. acanthias*) to be one of the most frequent shark species captured in the Mediterranean [30, 86, 87]. Whereas, its congener *S. blainville* is stated to be one of the most important species of the demersal assemblages in the eastern Ionian Sea [88], as well as throughout the basin, principally in its central-western part (eastern Corsica and southern Sicily) and the eastern Ionian and Aegean seas [87]. The species was found to be more abundant on the slope than on the shelf [89].

The reason for this replacement between those sharks could be related to taxonomic problems afflicting the *Squalus* genus in different areas of the Mediterranean [38, 84]. Indeed, recent studies [53] highlight that *S. acanthias* showed a limited geographic distribution in the past, suggesting an inaccurate classification of these two species [90].

To clarify the real presence of the *Squalus* species in the Mediterranean Sea, numerous scientific studies, such as MEDITS trawl survey, have been conducted.

	Method	Sex	VBGM parameters			Maturity age (years)	Maximum age (years)	Location
			L_∞ (cm)	K (years $^{-1}$)	T_0 (years)			
Squalus blainville	Dorsal fin spine	F	105.7	0.11	−1.1	7.4	19	Gulf of Gabès, Tunisia [66]
		M	91.1	0.14	−1.4	4.7	15	
	Vertebrate	F	117.9	0.10	−1.3	5.1	8	Strait of Sicily, Italy [29]
		M	96	0.13	−1.3	3.3	8	
	Dorsal fin spine	F	109.7	0.03	−5.58	17	28	eastern Mediterranean Sea [72]
		M	66.55	0.08	−3.35	11.3	22	
S. megalops	Dorsal fin spine	F	82.31	0.06	−3.89	15.38	29	Gulf of Gabès, Tunisia [77]
		M	68.55	0.08	−4.65	8.39	26	
S. acanthias	Dorsal fin spine	F	173.3	0.09	- 0.04	20.9	23	Adratic Sea [80]
		M	103.3	0.09	- 0.04	10.5	36	
	Dorsal fin spine	F	113.0	0.18	—	7.5	13	Adratic Sea [57]
		M	92.0	0.24	—	5.5	9	

Table 4.
Studies on Squalus blainville, S. megalops, *and* Squalus acanthias *age, carried out in the Mediterranean Sea (VBGM Von Bertallonffy; L∞: The asymptotic length at which growth is zero; K is the growth rate; T_0: Constant value).*

According to those studies, *S. acanthias* and *S. blainville* are mostly found in the northern part of the Mediterranean Sea, Adriatic, and Black Sea [25, 84, 85, 91].

Concerning *S. megalops*, it is recorded from the northern coasts of the Canary Islands, Morocco, and southern Spain (Malaga), but it is present mainly along the African coasts of Tunisia [41].

2.3 Catches

Fundamental problems in accurately identifying and classifying species have hampered the collection of robust biological and ecological data. This fact, together with erroneous reports of reported catches (usually lower than actual catches; [30, 92]) makes stock estimates for these fish difficult to assess. This situation is of particular concern in elasmobranchs taken as targets or as bycatch in several fisheries around the globe, as their conservative life history strategies make them extremely vulnerable to overexploitation [93].

Worldwide, Indonesia and Spain remain the top three shark catchers in the world [94]. In the Mediterranean Sea, countries, contributing more to the elasmobranch landings during the last years, are Tunisia and Libya, which contributed more than 70% of production. Tunisian landings did not show any distinguished variations from 1980 to 2015. Those from Libya appear for the first time in FAO statistics and seem to be significant. Turkey and Italy register a dramatic decrease in catch after being known to be the major elasmobranch-fishing countries within the Mediterranean, between 1980 and 2008. It should be noted that the Mediterranean landings of carcharhiniformes, the most represented group among the elasmobranchs and the most commercially fished, recorded a notable decrease [92]. The most commonly caught species are skates (Rajidae) and catsharks (*Scyliorhinus spp.* and *Galeus spp.*) [95, 96]. Different species of pelagic sharks, as well as eagle rays (Myliobatidae) and stingrays (Dasyatidae), are bycatch of pelagic and demersal fisheries [97, 98].

Squalidae, it represents one of the most commercially targeted families among elasmobranchs [24]. Capture production for Squalidae in the Mediterranean and the Black Sea during the last decades is illustrated in **Figure 1** [99].

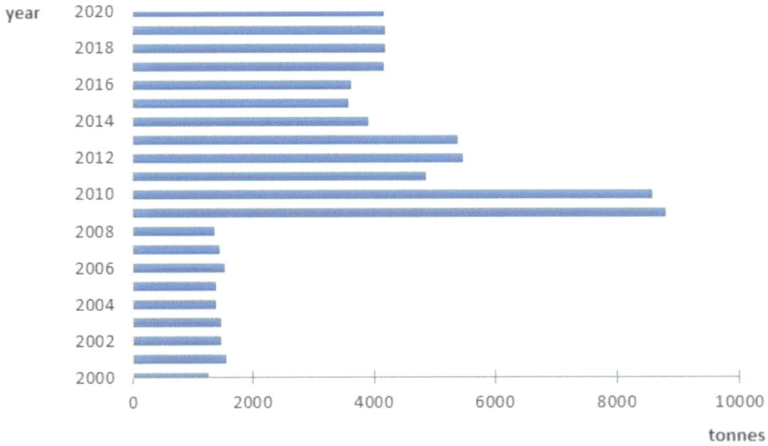

Figure 1.
Capture production for Squalidae in the Mediterranean and Black Sea during the last decades.

Indeed, several species belonging to this family are landed by up to 50 countries in direct fisheries or as bycatch [24]. The genus *Squalus* is highly represented in bycatch and several studies have focused on the mitigation of the fishery impact on this group [100–102].

In the Mediterranean Sea, *S. blainville* constitutes an important landing from bottom trawlers, longlines, and gill nets [103, 104]. Moreover, the presence of *S. megalops* and *S. blainville* was reported on African coasts of Tunisia in bycatch of bottom trawl and longline fisheries [41, 92].

Although longlines are considered selective, they bring several nontarget species, including *S. acanthias* [92]. In fact, a significant decline in the bycatch of the former species is perceived according to the fishermen's perception on the evolution of shark populations in the northern Catalan coast (north-western Mediterranean Sea) [105]. The low presence of the piked dogfish (Annex III of the Barcelona Convention) in the subregion's bycatch composition could confirm the recorded decrease in biomass of this species.

The bycatch of elasmobranchs is an issue of global concern, particularly in high-seas pelagic longline fisheries, where 25% of the catch is nontarget sharks and rays [106]. Thus, in order to keep the populations of these fishes in balance, incisive management programs are required to guarantee stability for the populations. Among these management measures, the elimination or at least the reduction of the bycatch cannot be missing.

2.4 Taxonomic status

Dependable data on species richness are crucial for any biodiversity study and conservation policies, even though it is every so often difficult to discriminate a species based on extremely similar morphological characters [107]. Therein, reliable species identification is the principal step for the application of conservation policies and maintainable exploitation of natural resources [108], all the more so considering the currently elevated biodiversity crisis induced by human activities [109].

Overall, sharks belonging to the *Squalus* genus exhibit conserved body morphology, making identification based entirely on morphological characters complicated, leading to misidentifications [110]. This complexity is amplified even similarly *via* the high overlap of morphological characters among species, as identification is often based on limited and insufficiently consistent characters, like the number of vertebrae and morphometric data [10, 16, 20, 38, 111, 112]. In fact, morphological and biological similarities among squalids have led to considerable confusion over their taxonomy [14]. Some taxonomic and nomenclatural problems affect the group of species in question. Excluding *S. acanthias*, easily recognizable thanks to its specific pattern, characterized by the presence of white spots on the back or narrowly round to acutely angular rear tips and inner margins of the pectoral fins, which permits an easier identification and discrimination from the other two species [84], and the correct identification of the other two species, which do show a very similar morphology, requires the observations of the dermal denticles, meristic features, or even genetic analysis.

Compared with *Squalus acanthias* and *S. asper*, a close similarity between *S. blainville* and *S. megalops* was pointed out [113, 114]. Moreover, despite that the relationships between the snout tip and nostril distance and the distance from the nostril to the preoral clefs were proposed as the best features for discriminating between species of the genus *Squalus*, and it is proved that they were of little use [115].

The taxonomic status of *S. blainville* is problematical as there are no extant types and the original description and figures do not correspond to any known species of

Squalus [42, 116]. Consequently, in a review of the Australian species of *Squalus*, *Squalus griffini* and *Squalus fernandinus* (Molina, 1782) were incorrectly synonymized with *S. blainville* [32]. However, in a review of Japanese *Squalus*, *S. blainville* was defined as a species with high dorsal fins and long dorsal-fin spines [116]. The same review revealed that *Squalus*, referring to *S. fernandinus* and *S. blainville* by some authors, had short dorsal-fin spines and were more similar to *S. mitsukurii* from Japan, and suggested that nominal *S. blainville* from New Zealand could be identical to *S. mitsukuri* [112, 117]. It was also noted that dogfish resembling *S. mitsukurii* occurred in Australia and New Zealand [31]. Outside its main distributional area (Mediterranean Sea and eastern Atlantic), *S. blainville* has also been recorded erroneously in Australia and New Zealand [112]. It was thought to be widespread in the Atlantic, Indian and Pacific Oceans [34, 117] as well as in Japan [116]. The distribution of *S. blainville* was restricted to the Mediterranean Sea and eastern Atlantic, and its records in the Pacific records were questioned [17]. The confusion is due largely to the poor original description and the lack of type material.

Taxonomic research on the genus *Squalus* conducted in the Tunisian waters revealed that *S. blainville* in this zone is not characterized by its high first dorsal fin and spine as was well thought-out by some authors, but rather it is a short-spine species [31, 116]. Comparing their data for *S. blainville* with the measurements of the same species in different regions, they noted that data generally agree despite there being some differences in the morphometrics between populations, and that Tunisian *S. blainville* specimens examined and the specimens studied by some authors [41–43, 114, 115, 118] have similar vertebral counts.

Recently, according to the studies conducted in other areas of the Mediterranean Sea and based on morphological and genetic (COI sequences) analyses, only one spurdog species, *S. blainville*, occurs in the Ionian, Libyan, Aegean Seas, and the Sardinian waters [38, 119].

These findings spotlighted the stretch of sea between Tunisia, southern Sicily, Malta, and Libya, known as the Strait of Sicily, as a more interesting area for spurdog species. The presence of *S. blainville* in the Maltese waters was assessed through the use of the DNA barcoding approach [84]. In the same region, some authors [37] collected and analyzed individuals belonging to the nominal *S. blainville* and genetically clustering within clade B [22], while three individuals were classified as *Squalus* sp. by the authors as clustering in the genetic clade C [22].

Regarding *S. megalops*, it is described for the first time in the Mediterranean in 1984 [43]; its occurrence is still debated and several scientific studies contributed to clarifying the real presence of this species in the area [21, 22, 37, 39, 41, 42, 84, 85, 119].

It was suggested that the southern Australian *S. megalops* might be endemic to Australia [91]. However, the tale can be even extra complex. Morphological research has proven that more than a single form of this species exists in Australian seas [28].

Recently, few research within the context of integrative taxonomy had been successfully accomplished in the genus *Squalus* aiming at the integration of new molecular taxonomy techniques to more classical morphological analyses with the purpose to make clear taxonomic ambiguities among some of the species [41, 84].

To highlight the taxonomic uncertainties in relation to the occurrence of *Squalus* species in the central Mediterranean Sea, a study including other morphometric characters and a molecular study was carried out to confirm that *S. megalops* occurs as a valid species in the Mediterranean Sea [41]. In fact, two species of spurdog of the genus *Squalus* occur in the Gulf of Gabès (southern Tunisia and central Mediterranean): *S. blainvillei* and *S. megalops cubensis* group. Morphometric and meristic data as well as genetic analyses (DNA inter-simple sequence repeat markers and molecular barcoding methods) support the assignation of this short-snout spurdog to *S. megalops*.

The Tunisian *S. megalops* species are consistent for characters typifying the "*megalops-cubensis*" group and fit the description of *S. megalops* from Australian waters [23], as well as the eastern Atlantic-Mediterranean [42] and Mediterranean waters [43]. Specimens described from other areas clearly agree with the Tunisian samples of *S. megalops* for most of the morphometric characters and had similar vertebral counts to those studied by other authors (Indo-Pacific, [118]; South Africa, [115]; Mediterranean coasts of Spain, [34]; east Atlantic, [42]; south western Australia, Queensland [23]).

Some authors stated that the number of chondrocranial lateral processes is the most important character to discriminate between *S. blainville* and *S. megalops* [34, 42], but they can be discriminated also based on other morphological features, such as teeth and dermic denticles morphology. These findings have been confirmed in the Gulf of Gabès through morphometric, meristic, and genetic analyses, suggesting that S. *megalops* could be even more common than *S. blainville* in these waters (**Figures 2** and **3**) [41]. In addition to the differences cited between those species, the study of their traits of the life history in the area revealed that they differ also in those terms [50, 51, 63, 66, 77].

Differently from the former study [41], a study on the intraspecific morphological variability in *S. blainville* did not identify diagnostic features (e.g., dermal denticles), which could effectively discriminate between *S. blainville* and *S. megalops*. The authors asserted that species identification based only on morphological characteristics can easily lead to taxonomic misidentifications, especially when multiple anatomical characters (e.g., skull and teeth morphology) are used [84].

To aid in clarifying the taxonomic status of *Squalus* species in the eastern Atlantic and Mediterranean, some authors assessed species diversity at the molecular level and evaluated the consistency in species identification in the region [22]. They confirmed unreliable species identification in the eastern Atlantic and Mediterranean *Squalus* and reinforced the need to revise the status of *S. megalops* and *S. mitsukurii* as they may include several distinct species distributed around the world. Nonetheless, the results provided by those authors suggest that a different species from the "true Australian" *S. megalops*, which remains unidentified, can occur in the eastern Atlantic and Mediterranean waters [22].

In any case, specimens of *S. megalops* for which the identification is considered feasible were rarely reported in the catches [38, 51]. Most of the catches of these species are recorded from the northern coasts of Canary Islands, Morocco, and southern Spain (Malaga). However, the real presence of *S. megalops* is still unclear not only for the Mediterranean Sea but also for the neighboring Atlantic area [18, 22] and some evidences confirm the inconsistency of the species identification keys to distinguish between the Atlantic and Mediterranean *Squalus*, concerning *S. blainville* and *S. megalops* [22].

Studies conducted in the Sardinian waters showed that morphological and genetic analysis revealed the presence only of *S. blainville* in the region, despite the observation of chondrocranial lateral processes initially allowing the investigated specimens to be subdivided into two groups. Indeed, the comparison of chondrocranial and body morphology of the specimens examined indicated that none of the considered measurements could differentiate the two squalid groups [38].

They noted that considering the brief half-life and fast replacement rate of the dermal denticles [120]. In fact, the different development stages of denticles observed

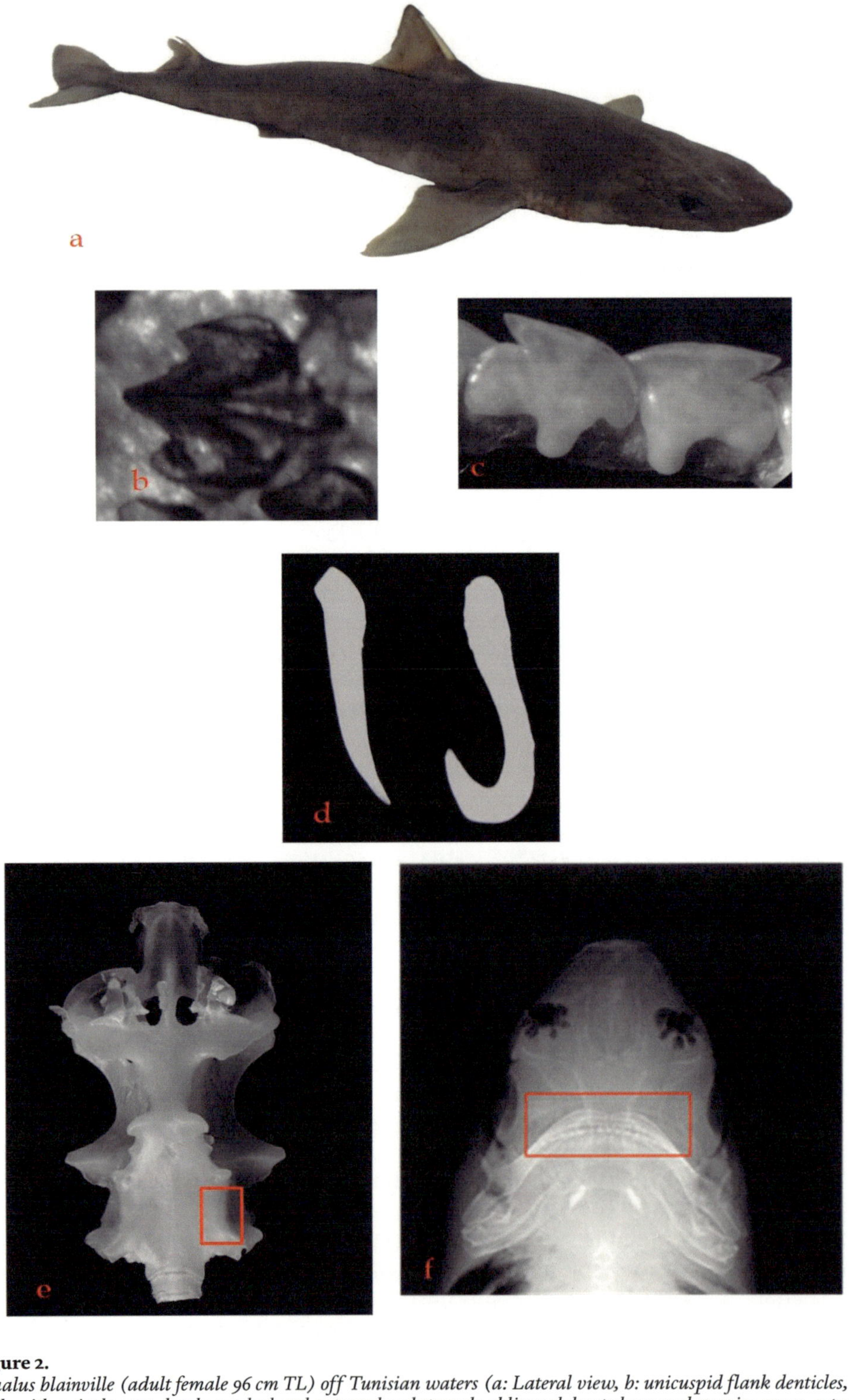

Figure 2.
Squalus blainville (adult female 96 cm TL) off Tunisian waters (a: Lateral view, b: unicuspid flank denticles, c: teeth with a single cusp deeply notched and outward end strongly oblique d: bent claws and massive spurs, e: two cartilaginous processes in the basal plate, f: sharpen palatoquadrate).

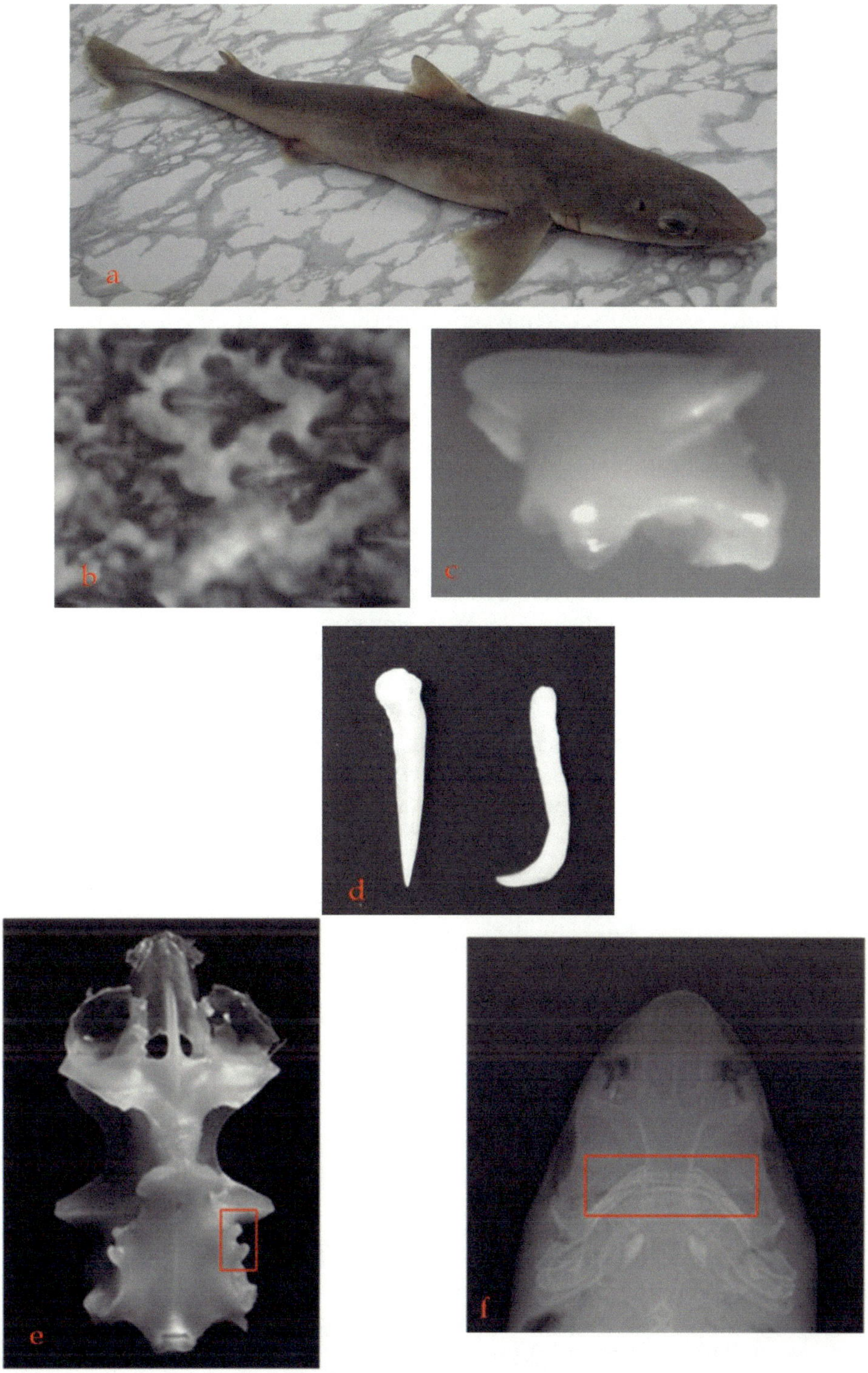

Figure 3.
Squalus megalops (adult female 76 cm TL) off Tunisian waters (a: lateral view, b: unicuspid flank denticles, c: teeth with a single cusp deeply notched and outward end directed strongly laterally, d: bent claws and massive spurs, e: two cartilaginous processes in the basal plate, f: sharpen palatoquadrate).

in the analyzed skin portion could explain this particular aspect [120]. Moreover, dermal denticles, teeth, and dorsal fin spines were reported as common diagnostic morphological structures, which could vary in shape with the ontogenetic development [121, 122]. Consequently, the morphology of the dermal denticles should be further investigated before it can be properly used as a suitable classification tool, as also suggested particularly for the genus *Squalus* [84]. The same authors stated that considering the finding of sporadic divergent sequences [22, 38, 41, 119] different from *S. blainville* and *S. acanthias* but also *S. megalops* from Australia, the occurrence of a third species in the Mediterranean (apart from *S. acanthias* and *S. blainville*) cannot be ruled out.

Similar efforts were undertaken recently for those species [39]; it was noted that *S. megalops* does not occur in the eastern Atlantic and Mediterranean waters and that individuals composing clade C [22] should be considered a new species that needs formal description and proper taxonomic assessment [22, 123]. Besides, the species described as short-snout spurdog by other authors was considered a rare species and an occasional visitor with high morphological similarity to the *S. megalops* and *S. blainville* but is genetically distinct from both [39].

According to some authors [32], molecular data alone do not replace traditional taxonomy in the delimitation of species [124]. This integrative approach has been used over the years and has proven to be quite effective in elasmobranchs [124, 125] and in other groups of organisms [125, 126]. However, because of the difficulty of morphologically defining *Squalus* species, many sequences available in genetic databases indicate misidentifications or identifications only at the genus or family levels, making them not very useful for molecular identification purposes.

On their part, some authors stated that despite *S. acanthias* is the type species of the genus and is one of the most easily distinguished species of *Squalus* some sequences of Mediterranean specimens originally recognized as *S. acanthias* clustered in clade B. This misidentification is surprising but reveals the confused state of *Squalus* taxonomy in the region [22].

These findings further support current inconsistencies in species identification within the genus *Squalus* and the need for an accurate redescription of *Squalus* species, especially in the Mediterranean Sea, to stabilize the systematic and facilitate specimens' identification.

3. Conclusions

The Mediterranean Sea represents some of the most intensively studied regions of the world's oceans; however, this wealth of information does not translate into a good understanding of the species diversity and raises additional concerns regarding accurate identification of elasmobranchs. This concerns, among others, the genus *Squalus,* which the taxonomic confusion on it, is intrinsically related to difficulties in morphologically separating congeners and to incessant applications of synonyms due to the lack of appropriate taxonomic scrutiny with disregard for detailed morphological assessments that are essential for understanding possible variations and defining species.

In conclusion, since the first comprehensive revision on the genus in Africa [115] and after over 40 years gap, it is clear thus that an integrative approach includes both morphological and genetic tools with rigorous participation of taxon experts in the systematics of elasmobranch fishes still need to be strengthened to reduce this "taxonomic obstacle" and to faster actions in conservation and management of target species that were formerly unknown. Therefore, the establishment of a coordinated international effort to implement a comprehensive and integrated taxonomic assessment of this genus which represents an irreplaceable component of the biodiversity of the area is welcomed.

Acknowledgements

The authors gratefully acknowledge the Regional Activity Center for Specially Protected Areas (SPA/RAC) for its contribution in funding the cost of editing this chapter.

Author details

Sondes Marouani*, Sami Karaa and Othman Jarboui
National Institute of Marine Science and Technology (Sfax Center), Sfax, Tunisia

*Address all correspondence to: sondes.marouani@instm.rnrt.tn

References

[1] Tortonese T. Osservazioni comparative intorno alla ittiofauna del Mediterraneo e dell'Atlantico occidentale (Florida e isole Bahamas). Natura. 1989;**1**:1-20

[2] Derrick DH, Cheok J, Dulvy NK. Spatially congruent sites of importance for global shark and ray biodiversity. PLoS One. 2020;**15**(7):e0235559. DOI: 10.1371/journal.pone.0235559

[3] Leone A, Puncher GN, Ferretti F, Sperone E, Tripepi S, Micarelli P, et al. Pliocene colonization of the Mediterranean by great White shark inferred from fossil records, historical jaws, phylogeographic and divergence time analyses. Journal of Biogeography. 2020;**47**(5):1119-1129. DOI: 10.1111/jbi.13794

[4] Bianchi CN, Morri C, Chiantore M, Montefalcone M, Parravicini V, Rovere A. Mediterranean Sea biodiversity between the legacy from the past and a future of change. In: Stambler N, editor. Life in the Mediterranean Sea: A Look at Habitat Changes. Vol. 1. New York: Nova Science Publishers, Inc; 2012. p. 55. ISBN: 978-1-61209-644-5

[5] Katsanevakis S, Coll M, Piroddi C, Steenbeek J. Ben Rais Lasram F, Zenetos a, Cardoso AC: Invading the Mediterranean Sea: Biodiversity patterns shaped by human activities. Frontiers in Marine Science. 2014;**1**:32. DOI: 10.3389/FMARS.2014.00032

[6] Serena F, Abella AJ, Bargnesi F, Barone M, Colloca F, Ferretti F, et al. Species diversity, taxonomy and distribution of Chondrichthyes in the Mediterranean and Black Sea. The European Zoological Journal. 2020;**87**:497-536. DOI: 10.1080/24750263.2020.1805518

[7] Dulvy NK, Allen DJ, Ralph GM, Walls RHL. The Conservation Status of Sharks, Rays and Chimaeras in the Mediterranean Sea [Brochure]. Malaga, Spain: IUCN; 2016. p. 14

[8] Dulvy NK, Fowler SL, Musick JA, Cavanagh RD, Kyne PM, Harrison LR, et al. White WT: Extinction risk and conservation of the world's sharks and rays. eLife. 2014;**3**:e00590. DOI: 10.7554/eLife.00590

[9] Otero M, Serena F, Gerovasileiou V, Barone M, Bo M, Arcos JM, et al. Identification Guide of Vulnerable Species Incidentally Caught in Mediterranean Fisheries. Malaga, Spain: IUCN; 2019. p. 204

[10] Menesini E. Ittiodontoliti Miocenici di Terra d'Otranto (Puglia). Palaeontographia Italica. 1968;**65**:1-61

[11] Ramírez-Amaro S, Ordines F, Puerto MÁ, García C, Ramon C, Terrasa B, et al. New morphological and molecular evidence confirm the presence of the Norwegian skate *Dipturus nidarosiensis* (storm, 1881) in the Mediterranean Sea and extend its distribution to the western basin. Mediterranean Marine Science. 2017;**18**(2):251-259. DOI: 10.12681/MMS.1950

[12] Zeeberg J, Corten A, Graaf E. Bycatch and release of pelagic megafauna in industrial trawler fisheries off Northwest Africa. Fisheries Research. 2006;**78**(2):186-195. DOI: 10.1016/j.fishres.2006.01.012

[13] Coelho R, Erzini K. Identification of deep water lantern sharks (Chondrichthyes: Etmopteridae) using morphometric data and multivariate analysis. Journal of the Marine Biological Association of the United Kingdom. 2008;**88**(1):199-204. DOI: 10.1017/S0025315408000271

[14] Myakov NA, Kondyurin VV. Dogfishes, *squalus* (Squalidae) of the Atlantic Ocean and comparative notes on the species of this genus from other regions. Journal of Ichthyology. 1986;**26**(6):1-18

[15] Ebert DA, Stehmann M. Sharks, Batoids, and Chimaeras of the North Atlantic. Vol. No. 7. Rome: FAO Species Catalogue of fishery Purpose; 2013. p. 523

[16] Ebert DA, Fowler S, Compagno LJV. Sharks of the World: A Fully Illustrated Guide. London: Wild Nature Press; 2013. p. 528

[17] Compagno LJV, Dando M, Fowler S. A Field Guide to the Sharks of the World. London: Harper-Collins; 2005. p. 496

[18] Last PR, White WT, Pogonosky JJ. Descriptions of the New Dogfishes of the Genus *Squalus* (Squaloidea: Squalidae). Hobart: CSIRO Marine and Atmospheric Research; 2007. p. 130. DOI: 10.4225/08/58615c42b11ce

[19] Viana ST, de Carvalho MR. Redescription of *Squalus acutipinnis* Regan, 1908, a valid species of spiny dogfish from southern Africa (Chondrichthyes: Squaliformes: Squalidae). Copeia. 2016;**104**:539-553. DOI: 10.1643/CI-14-217

[20] Viana ST, de Carvalho MR, Gomes UL. Taxonomy and morphology of species of the genus *squalus* Linnaeus, 1758 from the southwestern Atlantic Ocean (Chondrichthyes: Squaliformes: Squalidae). Zootaxa. 2016;**4133**:1-89. DOI: 10.11646/ZOOTAXA.4133.1.1

[21] Bellodi A, Porcu C, Cannas R, Cau A, Marongiu MF, Mulas A, et al. Life-history traits of the longnosed skate *Dipturus oxyrinchus*. Journal of Fish Biology. 2017;**90**(3):867-888. DOI: 10.1111/JFB.13205

[22] Veríssimo A, Zaera-Perez D, Leslie R, Iglesias SP, Seter B, Grigoriou P, et al. Molecular diversity and distribution of eastern Atlantic and Mediterranean dogfishes *squalus* highlight taxonomic issue in the genus. Zoologica scripta. 2017;**46**(4):414-428. DOI: 10.1111/zsc.12224

[23] Last PR, Edmunds M, Yearsley GK: Squalus crassispinus sp. nov., a new spurdog of the 'megalops-cubensis' group' from the eastern Indian Ocean. In Last PR, White WT, Pogonoski JJ editors. Descriptions of new dogfishes of the genus *Squalus* (Squaloidea:Squalidae). CSIRO Marine and Atmospheric Research Paper No 14, World Register of Marine Species. 2007; pp. 11-22

[24] Ebert DA, White WT, Goldman KJ, Compagno LJ, Daly-Engel TS, Ward RD. Resurrection and redescription of *Squalus suckleyi* (Girard, 1854) from the North Pacific, with comments on the *Squalus acanthias* subgroup (Squaliformes: Squalidae). Zootaxa. 2010;**2612**(1):22-40. DOI: 10.11646/zootaxa.2612.1.2

[25] White WT, Iglesias SP. *Squalus formosus*, a new species of spurdog shark (Squaliformes: Squalidae), from the western North Pacific Ocean. Journal of Fish Biology. 2011;**79**(4):954-968. DOI: 10.1111/j.1095-8649.2011.03068.x

[26] Borsa P, Shen KN, Arlyza IS, Hoareau TB. Multiple cryptic species in the blue-spotted maskray (Myliobatoidei: Dasyatidae: Neotrygon spp.): An update. Comptes rendus biologies. 2016;**339**(9-10):417-426. DOI: 10.1016/j.crvi.2016.07.004

[27] Borsa P, Arlyza IS, Hoareau TB, Shen KN. Diagnostic description and geographic distribution of four new cryptic species of the blue-spotted maskray species complex (Myliobatoidei:

Dasyatidae; Neotrygon spp.) based on DNA sequences. Journal of Oceanology and Limnology. 2018;**36**(3):827-841. DOI: 10.1007/s00343-018-7056-2

[28] Whitehead PJP, Bauchot ML, Hureau JC, Nielsen J, Tortonese E. Fishes of the North-Eastern Atlantic and the Mediterranean (FNAM). Paris: Unesco; 1984. p. 1473 ISBN 92-3-002308-6

[29] Cannizzaro L, Rizzo P, Levi D, Gancitano S. Age determination and growth of *Squalus blainvillei* (Risso, 1826). Fisheries Research. 1995; **23**(1-2):113-125. DOI: 10.1016/0165-7836(94)00333-R

[30] Serena F, Papacostantinou C, Relini G, Gil De Sola L, Bertrand J. Distribution and abundance of spiny dogfish in the Mediterranean Sea based on the Mediterranean international trawl surveys program. Biology and management of dogfish sharks. In: Gallucci VF, McFarlane GA, Bargmann GC, editors. American Fisheries Society. Maryland: Bethesda; 2009. pp. 139-149. DOI: 10.47886/9781934874073.ch13

[31] Compagno LJV. Squaliformes-Dogfish sharks. In: Fischer W, Nauen C, editors. FAO Species Catalogue, Volume 4. Sharks of the World: An Annotated and Illustrated Catalogue of Shark Species Known to Date. Part I. Hexanchiformes to Lamniformes. Vol. 125. Rome: FAO Fisheries Synopsis; 1984. pp. 111-113

[32] Garrick JAF. Studies on New Zealand Elasmobranchii. Part XII. The species of *squalus* from New Zealand and Australia; and a general account and key to the New Zealand Squaloidea. The Transactions and Proceedings of the Royal Society of New Zealand. 1960;**88**:519-557

[33] Garrick JAF, Paul LJ. *Heptranchias dakini* Whitley, 1931, a synonym of *H. perlo* (Bonnaterre, 1788), the sharpnosed sevengill or perlon shark, with notes on sexual dimorphism in this species. Zoological Publications from Victoria University of Wellington. 1971;**54**:1-14

[34] Bigelow HB. Schroeder WC 1957: A study of the sharks of the suborder Squaloidea. Bulletin of the Museum of Comparative Zoology. 1957;**117**:1-150

[35] Daly-Engel TS, Koch A, Anderson JM, Cotton CF, Rubbs RD. Description of a new deep-water dogfish shark from Hawaii, with comments on the *Squalus mitsukurii* species complex in the West Pacific. Zookeys. 2018;**798**:135-157. DOI: 10.3897/zookeys.798.28375

[36] Ziadi-Künzli F, Soliman T, Imai H, Sakurai M, Maeda K, Tachihara K. Re-evaluation of Deep-Sea dogfishes (genus *squalus*) in Japan using phylogenetic inference. Deep-Sea Research Part I: Oceanographic Research Papers. 2020;**160**:103261. DOI: 10.1016/j.dsr.2020.103261

[37] Vella A, Vella N, Schembri S. A molecular approach towards taxonomic identification of elasmobranch species from Maltese fisheries landings. Marine Genomics. 2017;**36**:17-23. DOI: 10.1016/j.margen.2017.08.008

[38] Bellodi A, Porcu C, Cau A, Marongiu MF, Melis R, Mulas A, et al. Investigation on the genus *squalus* in the Sardinian waters (Central-Western Mediterranean) with implications on its management. Mediterranean Marine Science. 2018;**19**:256-272. DOI: 10.12681/mms.15426

[39] Ferrari A, Di Crescenzo S, Cariani A, Crobe V, Benvenuto A, Piattoni F, et al. Puzzling over Spurdogs: Molecular taxonomy assessment of the *squalus* species in the strait of Sicily. The European Zoological Journal.

2021;**88**:181-190. DOI: 10.1080/24750263.2020.1849436

[40] Ward RD, Holmes BH, Zemlak TS, Smith PJ. DNA barcoding discriminates Spurdogs of the genus *squalus*. In: Last PR, White WT, Pogonoski JJ, editors. Descriptions of New Dogfishes of the Genus *Squalus* (Squaloidea: Squalidae). Hobart, Australia: CSIRO Marine and Atmospheric Research; 2007. pp. 117-130. Available from: http://hdl.handle.net/102.100.100/124477?index=1

[41] Marouani S, Chaâba R, Kadri H, Saidi B, Bouain A, Maltagliati F, et al. Taxonomic research on *Squalus megalops* (Macleay, 1881) and *Squalus blainvillei* (Risso, 1827) (Chondrichthyes: Squalidae) in Tunisian waters (Central Mediterranean Sea). Scientia Marina. 2012;**76**(1):97-109. DOI: 10.3989/scimar.03440.22B

[42] Muñoz-Chápuli R, Ramos F. Morphological comparison of *Squalus blainvillei* and *S. megalops* in the eastern Atlantic, with notes on the genus. Japanese Journal of Ichthyology. 1989;**36**(1):6-21. DOI: 10.1007/BF02905668

[43] Muñoz-Chápuli R, Ramos F, Garcia-Garrido CM. *Squalus megalops* (McLeay, 1882) en el Mediterraneo. Notas sobre su diagnosis, sistematica y distribucion. Butlleti de la Societat Catalana d'Ictiologia i Herpetologia. 1984;**9**:16-21

[44] King JR, McFarlane GA. Marine fish life history strategies: Applications to fishery management. Fisheries Management and Ecology. 2003;**10**:249-264. DOI: 10.1046/j.1365-2400.2003.00359.x

[45] Dell'Apa A, Bangley CW, Rulifson RA. Who let the dogfish out? A review of management and socio-economic aspects of spiny dogfish fisheries. Reviews in Fish Biology and Fisheries. 2015;**25**:273-295. DOI: 10.1007/s11160-014-9379-1

[46] Quignard JP. Recherche sur la biologie de *Squalus blainvillei*, (Risso, 1826). Travaux du Laboratoire de la Biologie Halieutique Université de Rennes. 1971;**5**:125-141

[47] Capapé C. Observations sur la sexualité, la reproduction et la fécondité de 8 sélaciens pleurotrêmes vivipares placentaires des côtes tunisiennes. Archives de l'institut Pasteur de Tunis. 1974;**51**:329-344

[48] Capapé C, Quignard JP. Recherche sur la biologie de *Squalus blainvillei*, (Risso, 1826) (Pisces, Squalidae) des côtes tunisiennes: relations taille-poids du corps, du foie et des gonades. Coefficients de condition. Rapports hépato et gonosomatique. Croissance embryonnaire. Archives de l'institut Pasteur Tunis. 1980;**57**(4):385-408

[49] Sion L, D'Onghia G, Tursi A, Mytilineou C. First data on distribution and biology of *Squalus blainvillei* (Risso, 1826) from the eastern Mediterranean Sea. The Journal of Northwest Atlantic Fishery Science. 2003;**31**:213-219. DOI: 10.2960/J.v31.a16

[50] Marouani S, Kadri H, Bradaï MN, Saïdi B, Bouaïn A. Size at maturity of *Squalus blainvillei* and *Squalus megalops* in the Gulf of Gabès (Tunisia). Rapports des réunions Commission internationale pour l'exploration scientifique de la Mer Méditerranée. 2013;**40**:466. DOI: 10.1029/2005PA001169

[51] Marouani S. Systématique et éco-biologie de deux requins *Squalus blainvillei* (Risso, 1827) et *Squalus megalops* (Macleay, 1881) (Chondrichtyens, Squalidae) du golfe de Gabès (Méditerranée centrale).

Tunisia: Faculté des sciences de Sfax: Université de Sfax; 2013

[52] Kousteni V, Megalofonou P. Reproductive biology and embryonic development of *Squalus blainvillei* in the eastern Mediterranean Sea. Scientia Marina. 2011;**75**(2):237-249. DOI: 10.3989/scimar.2011.75n2237

[53] Anastasopoulou A, Mytilineou C, Makantasi P, Smith CJ, Kavadas S, Konstantia EL, et al. Life history aspects of two species of the *squalus* genus in the eastern Ionian Sea. Journal of the Marine Biological Association of the United Kingdom. 2018;**98**(4):937-948. DOI: 10.1017/S0025315416001818

[54] Capapé C, Reynaud C. Maturity, reproductive cycle and fecundity of the spiny dogfish *Squalus acanthias* (Chondrichthyes: Squalidae) off the Languedocian coast (southern France, northern Mediterranean). Journal of the Marine Biological Association of the United Kingdom. 2011;**91**(8):1627-1635. DOI: 10.1017/S0025315411000270

[55] Chatzispyrou A, Megalofonou P. Sexual maturity, fecundity and embryonic development of the spiny dogfish, *Squalus acanthias*, in the eastern mediterranean sea. Journal of the Marine Biological Association of the United Kingdom. 2005;**85**(5):1155-1161. DOI: 10.1017/S0025315405012233

[56] Yigin CC, Ismen A. Reproductive biology of spiny dogfish *Squalus acanthias*, in the North Aegean Sea. Turkish Journal of Fisheries and Aquatic Science. 2013;**13**:169-177. DOI: 10.4194/1303-2712-v13_1_20

[57] Bargione G, Donato F, La Mesa M, et al. Life-history traits of the spiny dogfish *Squalus acanthias* in the Adriatic Sea. Scientific Reports. 2019;**9**:14317. DOI: 10.1038/s41598-019-50883-w

[58] Kousteni V, Karachle PK, Megalofonou P. Diet and trophic level of the longnose spurdog Squalus Blainville (Risso, 1826) in the deep waters of the Aegean Sea. Deep Sea Research Part I: Oceanographic Research Papers. 2017;**124**:93-102. DOI: 10.1016/j.dsr.2017.04.008

[59] Bonnici L, Bonello JJ, Schembri PJ. Diet and trophic level of the longnose spurdog, *Squalus Blainville* (Risso, 1826) in the 25-nautical mile fisheries management zone around the Maltese islands. Regional Studies in Marine Science. 2018;**19**:33-42. DOI: 10.1016/j.rsma.2018.03.001

[60] Bengil EGT. Biology and ecology of *squalus Blainville* (Risso, 1827) from the eastern Mediterranean. Thalassas. 2022;**124**:93-102. DOI: 10.1007/s41208-022-00482-w

[61] Barria C, Coll M, Navarro J. Unravelling the ecological role and trophic relationships of uncommon and threatened elasmobranchs in the western Mediterranean Sea (supplement). Marine Ecology Progress Series. 2015;**539**:225-240

[62] Güzin G, Demirel N. Evaluation of the comprehensive feeding strategy and trophic role of overexploited mesopredator species in the sea of Marmara (northeastern Mediterranean). Estuarine, Coastal and Shelf Science. 2021;**259**:107448. DOI: 10.1016/j.ecss.2021.107448

[63] Marouani S, Kadri H, Karaa S. MN Bradai: Feeding ecology of the piked *spurdog Squalus* megalops (Chondrichthyes: Squalidae) in the Gulf of Gabès (Central Mediterranean Sea). Marine and Freshwater Research. 2017;**69**(1):48-55. DOI: 10.1071/MF17018

[64] Merella P, Quetglas A, Alemany F, Carbonell A. Length-weight relationship

of fishes and cephalopods from the Balearic Islands (Western Mediterranean). Naga, The WorldFish Center. 1997;**20**(3/4):66-68

[65] Pallaoro A, Jardas I, Šantič M. Weight-length relationships for 11 chondrichthyan species in the eastern Adriatic Sea. Cybium. 2005;**29**(1):93-96

[66] Marouani S, Kadri H, Saidi B, Morize E, Bouain A, Bradai MN. Age, growth, longevity, natural mortality and maturity of the longnose spurdog *Squalus blainvillei* (Chondrichthyes: Squalidae) in the Gulf of Gabès (Central Mediterranean Sea). Cahiers de biologie marine. 2012;**53**:197-204. DOI: 10.21411/CBM.A.E6C632F8

[67] Ismen A, Ozen O, Altinagac U, Ozekinci U, Ayaz A. Weight–length relationships of 63 fish species in Saros Bay, Turkey. Journal of Applied Ichthyology. 2007;**23**:707-708. DOI: 10.1111/j.1439-0426.2007.00872.x

[68] Ismen A, Yigin C, Altinagac U, Ayaz A. Length–weight relationships for ten shark species from Saros Bay C (North Aegean Sea). Journal of Applied Ichthyology. 2009;**25**:109-112. DOI: 10.1111/j.1439-0426.2009.01263.x

[69] Bok TD, Gokturk D, Kahraman AE, Alicli TZ, Acun T, Ates C. Length-weight relationships of 34 fish species from the sea of Marmara, Turkey. Journal of Animal and Veterinary Advances. 2011;**10**:3037-3042. DOI: 10.3923/javaa.2011.3037.3042

[70] Güven O, Kebapçıoğlu T, Deval MC. Length–weight relationships of sharks in Antalya Bay, eastern Mediterranean. Journal of Applied Ichthyology. 2012;**28**:278-279. DOI: 10.1111/j.1439-0426.2011.01823.x

[71] Eronat EGT, Özaydın O. Length-weight relationship of cartilaginous fish species from Central Aegean Sea (Izmir Bay and Sığacık Bay). Ege Journal of Fisheries and Aquatic Sciences. 2014;**31**(3):119-125. DOI: 10.12714/egejfas.2014.31.3.01

[72] Kousteni V, Megalofonou P. Aging and life history traits of the longnose spiny dogfish in the Mediterranean Sea: New insights into conservation and management needs. Fisheries Research. 2015;**168**:6-19. DOI: 10.1016/j.fishres.2015.03.014

[73] Lteif M, Mouawad R, Jemaa S, Khalaf G, Lenfant P, Verdoit-Jarraya M. The length-weight relationships of three sharks and five batoids in the Lebanese marine waters, eastern Mediterranean. Egyptian Journal of Aquatic Research. 2016;**42**:475-477

[74] Öztekin A, Özekinci U, Daban İB. Lengthweight relationships of 26 fish species caught by longline from the Gallipoli peninsula, Turkey (northern Aegean Sea). Cahiers de Biologie Marine. 2016;**57**:335-342

[75] Grace E, Bengil T. Biology and ecology of *squalus Blainville* (Risso, 1827) from the eastern Mediterranean. Thalassas: An International Journal of Marine Sciences. 2022. DOI: 10.1007/s41208-022-00482-w [In press]

[76] Falsone F, Geraci ML, Scannella D, Gancitano V, Di Maio F, Sardo G, et al. Length-weight relationships of 52 species from the south of Sicily (Central Mediterranean Sea). Fishes. 2022;**7**(2):92. DOI: 10.3390/fishes7020092

[77] Marouani S, Kadri H, Bradaï MN. First data on the age, growth and sexual maturity of the piked spurdog, *Squalus megalops* (Chondrichthyes: Squalidae), in the Gulf of Gabès (Central Mediterranean Sea). Marine and Freshwater Research. 2015;**67**(5):578-585. DOI: 10.1071/MF14356

[78] Filiz H, Mater S. Preliminary study on length-weight relationships for seven elasmobranch species from North Aegean Sea. Turkey Journal of Fisheries & Aquatic Sciences. 2002;**19**(3-4):401-409

[79] Filiz H. G bilge : Length–weight relationships of 24 fish species from the North Aegean Sea. Turkey. Journal of Applied Ichthyology. 2004;**20**(5):431-432. DOI: 10.1111/j.1439-0426.2004.00582.x

[80] Gracan R, Zavodnik D, Krstinić P, Lazar B. Feeding ecology and trophic segregation of two sympatric mesopredatory sharks in the heavily exploited coastal ecosystem of the Adriatic Sea. Journal of Fish Biology. 2016;**90**(1):167-184. DOI: 10.1111/jfb.13158

[81] Jones TS, Ugland KI. Reproduction of female spiny dogfish, Squalus acanthias, in the Oslofjord. Fishery Bulletin. 2001;**99**:685-690

[82] Sims CA. Implications of rational inattention. Journal of Monetary Economics. 2003;**50**(3):665-690. DOI: 10.1016/S0304-3932(03)00029-1

[83] Bat L, Erdem Y, Ustaoğlu S, Yardım Ö, Satılmıs HH. A study on the fishes of the Central Black Sea coast of Turkey. Journal of the Black Sea/Mediterranean Environment. 2005;**11**:281-296

[84] Bonello J, Bonnici L, Ferrari A, Cariani A, Schembri P. Not all that clear cut: Intraspecific morphological variability in *Squalus Blainville* (Risso, 1827) and implications for identification of the species. Journal of the Marine Biological Association of the United Kingdom. 2016;**96**(8):1585-1596. DOI: 10.1017/S0025315415001915

[85] Giarruso ME. Indagini morfologiche e genetiche sul genere Squalus nello stretto di Sicilia. Corso di laurea magistrale in biologia marina [thesis]. Italy: Tesi di laurea in struttura e connettività delle popolazioni marine; 2019. p. 31

[86] Baino R, Serena F, Ragonese S, Rey J, Rinelli P. Catch composition and abundance of elasmobranchs based on the MEDITS program. Rapport de la Commision Internationale pour l'Exploration Scientifique de la Méditerranée. 2001;**36**:234

[87] Damalas D, Vassilopoulou V. Chondrichthyan by-catch and discards in the demersal trawl fishery of the Central Aegean Sea (eastern Mediterranean). Fisheries Research. 2011;**108**(1):142-152. DOI: 10.1016/j.fishres.2010.12.012

[88] Tsagarakis K, Mytilineou C, Haralabous J, Lorance P, Politou CY, Dokos J. Mesoscale spatio-temporal dynamics of demersal assemblages of the eastern Ionian Sea in relationship with natural and fisheries factors. Aquatic Living Resources. 2013;**26**:381-397. DOI: 10.1051/alr/2013067

[89] Marongiu MF, Porcu C, Bellodi A, Cannas R, Cau A, Cuccu D, et al. Temporal dynamics of demersal chondrichthyan species in the central western Mediterranean Sea: The case study in Sardinia Island. Fisheries Research. 2017;**193**:81-94. DOI: 10.1016/j.fishres.2017.04.001

[90] Follesa MC, Marongiu MF, Zupa W, Bellodi A, Cau A, Cannas R, et al. Spatial variability of Chondrichthyes in the northern Mediterranean. Scientia Marina. 2019;**83**(S1):81-100. DOI: ORG/10.3989/SCIMAR.04998.23A

[91] Last PR. Stevens JD: Sharks and Rays of Australia. Melbourne, Australia: CSIRO; 1994. p. 513. DOI: 10.1017/S0025315400047858

[92] Bradaï NM, Saidi B, Enajjar S. Overview on Mediterranean Shark's

fisheries: Impact on the biodiversity. In: Türkoglu M, Önal U, Ismen A, editors. Marine Ecology - Biotic and Abiotic Interactions. London: Intechopen; 2018. pp. 211-230. DOI: 10.5772/INTECHOPEN.74923

[93] Musick JA, Burgess GH, Cailliet GM, Camhi M, Fordham SV. Management of sharks and their relatives(Elasmobranchii). Fisheries. 2000;**25**(3):913. DOI: 10.1577/15488446(2000)025<0009:MOSATR>2.0.CO;2

[94] FAO: Thc Statc of World Fisheries and Aquaculture. Sustainability in Action. Vol. 2020. Rome: FAO; 2020. p. 244. DOI: 10.4060/ca9229en

[95] Walker JMT, Joan MT, Wilkins AS, Dallaire JR, San dler HM, Hoover-Dempsey KV. Parental involvement: Model revision through scale development. The Elementary School Journal. 2005;**106**(2):85-104. DOI: 10.1086/499193

[96] Cashion MS, Bailly N, Pauly D. Official catch data underrepresent shark and ray taxa caught in Mediterranean and Black Sea fisheries. Marine Policy. 2019;**105**(7437):1-9. DOI: 10.1016/j.marpol.2019.02.041

[97] Damalas D, Megalofonou P. Occurrences of large sharks in the open waters of the southeastern Mediterranean Sea. Journal of Natural History. 2012;**46**:43-44. DOI: 10.1080/00222933.2012.716864

[98] Echwikhi K, Saidi B, Bradai MN, Bouain A. Preliminary data on elasmobranch gillnet fishery in the Gulf of Gabès Tunisia. Journal of Applied Ichthyology. 2013;**29**:1080-1085. DOI: 10.1111/JAI.12022

[99] GFCM (Mediterranean and Black Sea) Capture Production Quantity (1970 - 2020) https://www.fao.org/fishery/statisticsquery/en/gfcm_capture/gfcm_capture_quantity

[100] Mandelman JW, Farrington MA. The physiological status and mortality associated with otter-trawl capture, transport, and captivity of an exploited elasmobranch, *Squalus acanthias*. ICES Journal of Marine Science. 2007;**64**(1):122-130. DOI: 10.1093/icesjms/fsl003

[101] Stoner A, Kaimmer SM. Reducing elasmobranch bycatch: Laboratory investigation of rare earth metal and magnetic deterrents with spiny dogfish and Pacific halibut. Fisheries Research. 2008;**92**(2):162-168. DOI: 10.1016/j.fishres.2008.01.004

[102] Tallack SML, Mandelman JW. Do rare-earth metals deter spiny dogfish? A feasibility study on the use of electropositive "mischmetal" to reduce the bycatch of *Squalus acanthias* by hook gear in the Gulf of Maine. ICES Journal of Marine Science. 2009;**66**(2):315-322. DOI: 10.1093/icesjms/fsn215

[103] Séret B, Serena F. Distribution and abundance of *Squalus acanthias* and *Squalus blainvillei* in the Mediterranean Sea based on the Mediterranean international trawl survey program (MEDITS). In: The Mediterranean Chondrichthyan Fishes (Sharks, Rays, Skates and Chimaeras): Status and Priorities for Conservation. Tunis, Tunisia: United National Environment Programme (UNEP) Regional Activity Centres/Special Protected Areas (RAC/SP) Final Report; 2002

[104] Marongiu MF, Porcu C, Bellodi A, Cannas R, Carbonara P, Cau A, et al. Abundance, distribution and reproduction of the data-deficient species (*Squalus Blainville*) around Sardinia Island (central western

Mediterranean Sea) as a contribution to its conservation. Marine Freshwater Resources. 2020;**72**(1):118-130. DOI: 10.1071/MF19372

[105] Nuez I, Gazo M. Cardona : A closer look at the bycatch of medium-sized and large sharks in the northern Catalan coast (north-western Mediterranean Sea): Evidence of an ongoing decline? Aquatic Conservation: Marine and Freshwater Ecosystems. 2021;**31**(9):2369-2380. DOI: 10.1002/aqc.3651

[106] Mandelman JW, Cooper PW, Werner TB, Lagueux KM. Shark bycatch and depredation in the U.S. Atlantic pelagic longline fishery. Reviews in Fish Biology and Fisheries. 2008;**18**:427-442. DOI: 10.1007/s11160-008-9084-z

[107] Last PR, White WT, Pogonoski JJ, Gledhill DC, Yearsley GK, Ward RD: Application of a rapid taxonomic approach to the genus *squalus*. In: Last PR, White WT, Pogonoski JJ editors. Descriptions of New Dogfishes of the Genus *Squalus* (Squaloidea: Squalidae). Hobart, Tas.: CSIRO Marine and Atmospheric Research; 2007. p. 1-10. DOI: http://hdl.handle.net/102.100.100/124361?index=1

[108] Díaz-Jaimes P, Bayona-Vásquez NJ, Escatel-Luna E, Uribe-Alcocer M, Pecoraro C, Adams DH, et al. Population genetic divergence of Bonnethead sharks *Sphyrna tiburo* in the Western North Atlantic: Implications for conservation. Aquatic Conservation, Marine and Freshwater Ecosystems. 2021;**31**(1):83-98. DOI: 10.1002/aqc.3434

[109] Ceballos G, Ehrlich PR, Barnosky AD, García A, Pringle RM, Palmer TM. Accelerated modern human–induced species losses: Entering the sixth mass extinction. Science Adverse. 2015;**1**(5):e1400253. DOI: 10.1126/sciadv.1400253

[110] Geraci ML, Ragonese S, Norrito G, Scannella D, Falsone F, Vitale S: A tale on the demersal and bottom dwelling Chondrichthyes in the south of Sicily through 20 years of scientific survey. In: Filho LFDSR, de Luna Sales JR, editors. Chondrichthyes - Multidisciplinary Approach [Internet]. London: IntechOpen. 2017. p. 13-37. DOI: 10.5772/intechopen.69333

[111] Pfleger MO, Grubbs RD, Cotton CF, Daly-Engel TS. *Squalus clarkae Sp.* Nov., a new dogfish shark from the Northwest Atlantic and Gulf of Mexico, with comments on the *Squalus mitsukurii* species complex. Zootaxa. 2018;**4444**:101-119. DOI: 10.11646/zootaxa.4444.2.1

[112] Ariza AA, Adachi AMCL, Roque P, Hazin FHV, Vianna M, Rotundo MM, et al. DNA barcoding and species delimitation for dogfish sharks belonging to the *squalus* genus (Squaliformes: Squalidae). Diversity. 2022;**14**:544. DOI: 10.3390/d14070544

[113] Ledoux JC. Affinités et origines du *Squalus blainvillei* de Méditerranée. Journees Ichthologique, C.I.E.S.M. 1970;**3**:65-69

[114] Merrett NF. A new shark of the genus *squalus* (Squalidae, Squaloidea) from the equatorial western Indian Ocean, with notes on the *Squalus blainvillei*. Journal of Zoology. 1973;**171**:93-110. DOI: 10.1111/j.1469-7998.1973.tb07518.x

[115] Bass AJ, D'Aubrey JD, Kistnasamy N. Sharks of the east coast of southern Africa. VI the families Oxynotidae, Squalidae, Dalatiidae, and Echinorhinidae, Investigation. Investigational Report Institute of Oceanographic Sciences. 1976;**45**:1-103

[116] Chen C, Taniuchi T. Nose: Blainville's dogfish, *Squalus Blainville*,

from Japan, with notes on *S. mitsukurii* and *S. japonicus*. Japanese Journal of Ichthyology. 1979;**26**:26-42

[117] Bigelow HB, Schroeder WC: Fishes of the western North Atlantic, part I: Lancelets, cyclostomes, sharks. In Tee–Van J, Breder CM, Hildebrand SF, Parr AE, and Schroeder WC, Editors. Fishes of the Western North Atlantic. Part 1. Sears Foundation for Marine Research, Yale University, New Haven; 1948. P 59-576) DOI: 10.1126/science.111.2874.95

[118] Springer VG, Garrick JAF. A survey of vertebral numbers in sharks. Proceedings of the United States National Museum. 1964;**116**:73-96. DOI: 10.5479/si.00963801.116-3496.73

[119] Kousteni V, Kasapidis P, Kotoulas G, Megalofonou P. Evidence of high genetic connectivity for the longnose spurdog *Squalus Blainville* in the Mediterranean Sea. Mediterranean Marine Science. 2016;**17**(2):371-383. DOI: 10.12 681/mms.1222

[120] Kemp NE. Integumentary system and teeth. In: Hamlett WC, editor. Sharks, Skates and Rays: The Biology of Elasmobranch Fishes. Chapter 2. The John Hopkins University Press; 1999. pp. 43-68

[121] White WT, Ebert DA, Naylor GJP, Ho HC, Clerkin P, Veríssimo A, et al. Revision of the genus *Centrophorus* (Squaliformes: Centrophoridae): Part 1—Redescription of *Centrophorus granulosus* (Bloch & Schneider), a senior synonym of *C. acus Garman* and *C. niaukang Teng*. Zootaxa. 2013;**3752**(1):35. DOI: 10.11646/ZOOTAXA.3752.1.5

[122] Veríssimo A, Cotton CF, Buch RH, Guallart J, Burgess GH. Species diversity of the deep-water gulper sharks (Squaliformes: Centrophoridae: *Centrophorus*) in North Atlantic waters - current status and taxonomic issues. Zoological Journal of the Linnean Society. 2014;**172**(4):803-830. DOI: 10.1111/zoj.12194

[123] Last PR, Stevens JD. Sharks and Rays of Australia. 2nd ed. Collingwood, Australia: CSIRO; 2009. p. 644

[124] Schlick-Steiner BC, Steiner FM, Seifert B, Stauffer C, Christian E, Crozier RH. Integrative taxonomy: A multisource approach to exploring biodiversity. Annual Review of Entomology. 2010;**55**:421-438. DOI: 10.1146/annurev-ento-112408-085432

[125] Gabbanelli V. Díaz de Astarloa JM, Gonzalez-Castro M, Vazquez DM, Mabragaña E: Almost a century of oblivion: Integrative taxonomy allows the resurrection of the longnose skate Zearaja brevicaudata (Marini, 1933) (Rajiformes; Rajidae). Comptes Rendus Biologies. 2018;**341**:454-470. DOI: 10.1016/j.crvi.2018.10.002

[126] García-Melo JE, Oliveira C, Silva GJDC, Ochoa-Orrego LE, Pereira LHG, Maldonado-Ocampo JA. Species delimitation of Neotropical characins (Stevardiinae): Implications for taxonomy of complex groups. PLoS One. 2019;**14**:e0216786. DOI: 10.1371/journal.pone.0216786

Section 2

Sharks Fisheries

Chapter 3

Shark Fishing in Ghana: What We Ought to Know

Samuel K.K. Amponsah, Rachael Ackah, William Dzisenu Amekor, Asiedu Berchie and Andrews Apraku

Abstract

The main objective of the study was to assess the abundance and distribution of sharks in Ghana's coastal area. Samples were collected daily from the three sampling sites from April – to December 2021. The data obtained from this study were assessed for species abundance and composition, ecological indices; species diversity, and exploitation status using PAST and Microsoft Excel. A total of eight species were recorded with *Prionace glauca* (68%) as the dominant species and *Carcharhinus leucas* (1%) as the least dominant species. The mean species richness index (SRI) of 0.9 indicated infinite diversity of the species. The mean species evenness index (SEI) of 0.5 showed an evenly distributed species. The mean species dominance index (SDI) of 0.4 implied that the habitat was not dominated by only one or two particular species. The mean Shannon Weiner index (SWI) was less than 3, which is suggestive of pollution and habitat degradation. The exploitation status indicated that the stock of the shark species is in healthy condition. The month with the higher index was October which is possibly the aftermath of the close season and the major upwelling season. Extending the period of the closed season and reducing fishing effort are recommended.

Keywords: Ghana, exploitation status, diversity, sharks, diversity indices

1. Introduction

Sharks are a small, evolutionarily conservative group, comprising approximately more than 1200 species that have functioned successfully in diverse ecosystems for 400 million years [1]. Sharks perform numerous essential functions, both ecologically and economically. For instance, ecologically, shark species can act as both apex predators thereby securing the food web both directly by regulating prey dynamics through predation and indirectly by modifying prey behaviour and function as well as macro predators in line with a diverse group of reefs fish [2]. Although shark species are evolutionary successful, yet still some sharks are threatened with extinction because of human activity, climate change, pollution, and shark fin marketing as about 73 million sharks are killed every year according to an analysis of the Hong Kong shark fin trade [3]. Economically, shark fisheries sustain a substantial number of coastal communities' livelihoods [4]. Shark population declines can have unforeseeable implications, such as the collapse of significant fisheries. Many shark species have experienced significant population reductions as a result of the harmful effects of both target and non-target shark fishing [5]. Because sharks are especially sensitive to overfishing, there is evidence suggesting that some populations of large sharks have declined regionally by 90% or more [6]. Also, according to global reports, the shark population has declined by 70–80 percent [3]. Worldwide, shark populations are in grave peril. Despite questions over the sustainability of shark fisheries globally, information on world shark catches is often inadequate and regionally incomplete [7]. Consequently, few information is known on and landings of elasmobranchs, and species-level data are almost non-existent until recently, little attention has been afforded to the management of elasmobranch resources either by fisheries scientists and managers or by conservationists [8]. As a result of this, the status and trend of sharks globally are not appealing, and thus difficult to obtain the exact trend of the shark population globally. Apparently, over the last 60 years, shark catches by industrial, artisanal, and sport fisheries have increased around the world and sharks are now among the most threatened marine animals [9].

In Ghana, the shark fishery first developed around 1974 [10]. Shark fishing in West Africa has been undertaken as a commercial activity since the beginning of the 19th century, developed as a result of the growing demand for shark oil for lighting purposes [11]. Many fishers and traders generated between 80 and 100% of their income from shark fisheries [3]. In Ghana, much attention has been shifted to Shark fishing since their demand in the world is higher and hence valuable. One major threat to shark fisheries in Ghana is the unregulated of species mainly because they are caught as bycatch, and the meat is mostly used as bait for higher commercial species such as tuna, anchovies, and mackerels. Since the late 1950s, shark landings have been increasingly erratic in Ghana, peaking at 11,478 tons, in the last decade, the total reported shark catches fluctuated considerably. The catch peaked at 10,000 tons in 2013 and dropped to 8152 tons in 2015. In Ghana, shark fishing activities are not regulated mainly because the species are caught as bycatch, and the meat is mostly used as bait for higher commercial species such as tuna, anchovies, and mackerels, as today, shark products can reach expensive prices in Ghanaian markets [3, 12]. In Ghana, studies on sharks' distribution, abundance and species composition include the study by on 'Fishing for survival in the Western region of Ghana, [12] studies on 'Detection of illegal, unreported, and unregulated (IUU) fishing in sharks using barcoding in the Greater Region of Ghana. Another study was carried out by [3] on' Species composition, seasonality, and biological characteristics of Western Ghana's elasmobranch fishery. These studies however did not assess the maturity composition and the exploitation status of shark species. This study aimed to assess the diversity of shark species, estimate the diversity indices of shark species, determine the maturity composition of shark species, and determine the exploitation status of shark species on the coast of Ghana which could be used in aiding the sustainable management of fishery resources.

2. Materials and methods

2.1 Study area

The study was conducted in three coastal communities in the Western Region namely Axim and, Dixcove and Greater Accra Region namely Tema in Ghana (**Figure 1**). These three communities are the hotspots of shark fisheries in Ghana. Dixcove is located in Ghana's Western Region (N 04.79368°, W 01.94612°), Axim is

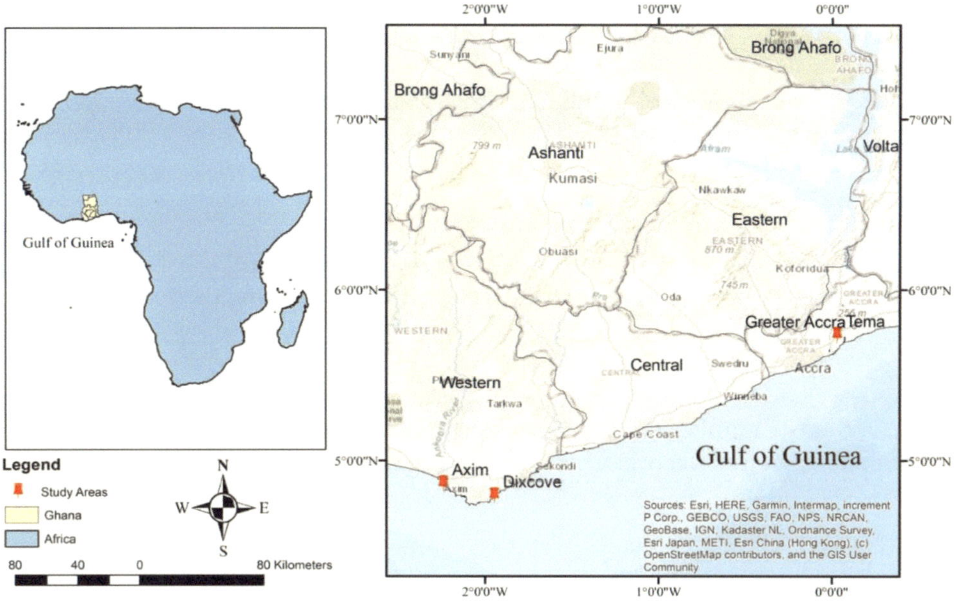

Figure 1.
Map showing the study sites.

located in the Nzema East district (N 04.8665° N, N 04.2409° W) and the Tema fishing community is situated at the Tema Newtown within the Tema Metropolitan Assembly.

2.2 Data collection

Samples were collected daily from three (3) sampling sites, i.e., from April to June and from August to December 2021. Majority of the fishers in these study areas use drift gill nets (DGN) in fishing for shark's species. However, in July there was no sampling due to the one (1) month banned on fishing activities in Ghana. On the field, the fish catch was identified to species level according to [13]. The length of fish samples was measured using a tape measure. Mostly the colour and the local names (mostly asked by the local fishermen) of sharks are noted down.

2.3 Species relative abundance

Relative species abundance is how rare or common a species is relative to other species in a defined location. This was expressed in percentage, using the expression below [14]:

$$\frac{\text{Number of individuals of species}}{\text{Total number of individuals}} * 100. \qquad (1)$$

2.4 Diversity indices

2.4.1 Shannon–Weiner index

Shannon Weiner's diversity index according to [15] considers both the number of species and the distribution of individuals among species. This index was calculated using the formula [16]:

$$H = -\sum\left[\left(\frac{m}{N}\right) \times \ln\left(\frac{ni}{N}\right)\right] \tag{2}$$

where ni is the number of individuals in species **i** and **N** is the total number of individuals in the community.

2.4.2 Species richness

This is the number of different species represented in an ecological community. Margalef index (d) was to measure the species richness by using the formula [17];

$$d = \frac{(S-1)}{\ln N} \tag{3}$$

Where S is the number of different species represented in the sample and N is the total number of individual organisms in your sample.

2.4.3 Species dominance

Simpson diversity index is the measure of diversity that takes into account the number of species present as well as the relative abundance of each species. This index was estimated using the formula [18]:

$$D = \sum n/N^2 \tag{4}$$

The value of D ranges between 0 and 1. With this index, 1 represents infinite diversity and 0, no diversity.

2.4.4 Species evenness

This index refers to how close in number each species in an environment is. Pielou's evenness index was used to calculate the evenness of the fish species in the sample. This was estimated using the formula [19]:

$$J = \frac{H^I}{H^I max} \tag{5}$$

where H' is the number derived from the Shannon diversity index and H'_{max} is the maximum possible value of H' (if every species was equally likely).

2.5 Exploitation rates

A rapid evaluation of the exploitation status of the most frequently landed species in the artisanal fishery was performed from a simple length-frequency framework developed by [20] for data-deficient fisheries and provides a first approximation of population parameters in these fisheries. The length-frequency framework uses empirical relationships between the asymptotic length (L_∞, cm), the mean length at first maturity (L_m, cm), and the length corresponding to the mean age in years at maximum possible yield per recruit, known as the optimum length (L_{opt}, cm). The following empirical relationships from [20] were used to estimate L_∞, L_m, and L_{opt}:

Asymptotic length (L_∞) was estimated from the maximum observed length (L_{max}) using the equation [20]:

$$\log L\infty = 0.044 + 0.9841 \log L_{max}. \quad (6)$$

Length at first maturity (L_m) was estimated from L_∞, as follows [20]:

$$\log Lm = 0.8979 \log L\infty - 0.0782 \quad (7)$$

where the standard error (SE) provides a measure of variability around the regression coefficient.

Length at maximum possible yield per recruit (L_{opt}) was estimated from L_m for unsexed fish, as follows [20]:

$$\log Lopt = 1.053 \log Lm - 0.0565. \quad (8)$$

The derived growth parameters (L_∞, L_m, and L_{opt}) were then indicated on the length-frequency distributions of the species to evaluate the exploitation status and sustainability of sharks caught in the artisanal fishery.

2.6 Data analysis

Descriptive statistics such as the mean, median and range were estimated using the length frequency distribution, and species diversity indices data. Frequency statistics were applied in showing the number of species obtained in each sampling area with other species. The statistical packages used for the study were Microsoft Excel and Palaeontological statistics software (PAST) Version 4. The Microsoft Excel Tool was used in estimating the descriptive statistic of the recorded length data of the species which involved the mean, median and range. The species diversity indices were done using the PAST V4.0 software.

3. Results

3.1 Species composition

Overall, eight (8) sharks' species were recorded during the study period (April 2021 to Dec 2021) as shown in **Figure 2**. They were *Carcharhinus leucas, Carcharias taurus, Isurus oxyrinchus, Prionace glauca, Sphyrna lewini, Carcharhinus brevipinna, Alopias supercilliosus,* and *Rhizoprionodon acutus.* The highest number of shark species were recorded in August and October (i.e. 7 species) and the least number of shark species were recorded in April, June and December (i.e. 4 species). The month of July was a close season so therefore there were no sharks landing (**Table 1**).

3.2 Species abundance

Figure 2 shows the overall relative abundance of species recorded from shark species landed in all the three selected communities from April 2021 to December 2021. From the Eight (8) species of sharks were observed with *Prionace glauca* as the most dominant species (68%) followed by *Rhizoprionodon acutus* (14%) as the second

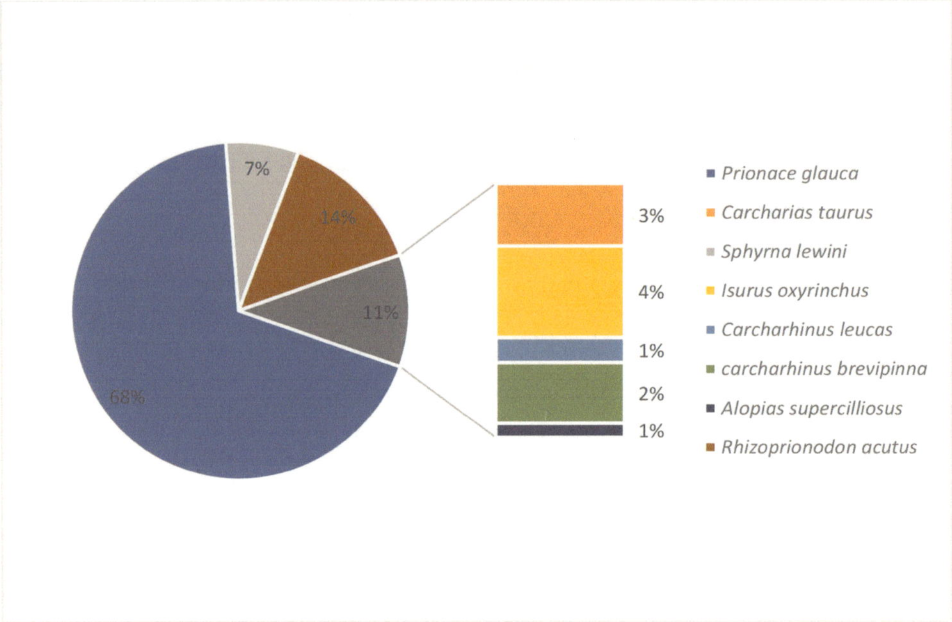

Figure 2.
Species abundance and composition from the study.

Species	April	May	Jun	Jul	Aug	Sep	Oct	Nov	Dec
Carcharhinus leucas	√	√	—	—	√	—	√	—	—
Carcharias taurus	√	√	—	—	—	—	√	√	—
Isurus oxyrinchus	—	√	√	—	√	√	√	√	—
Prionace glauca	√	√	√	—	√	√	√	√	√
Sphyrna lewini	√	√	√	—	√	√	√	√	√
Carcharhinus brevipinna	—	—	√	—	√	√	√	—	√
Alopias supercilliosus	—	—	—	—	√	√	—	—	—
Rhizoprionodon acutus	—	—	—	—	√	√	√	√	√
Total	4	5	4	—	7	6	7	5	4
NB: '√' means Present '-' means absent.									

Table 1.
Temporal composition of shark species from the study.

dominant and both *Carcharhinus leucas* (1%) and *Alopias supercilliosus* (2%) as the fewer dominate species.

3.3 Maturity composition

The composition of adults and juveniles' shark's species are showed in **Figure 3** where juveniles formed the lowest composition of all the species recorded except for

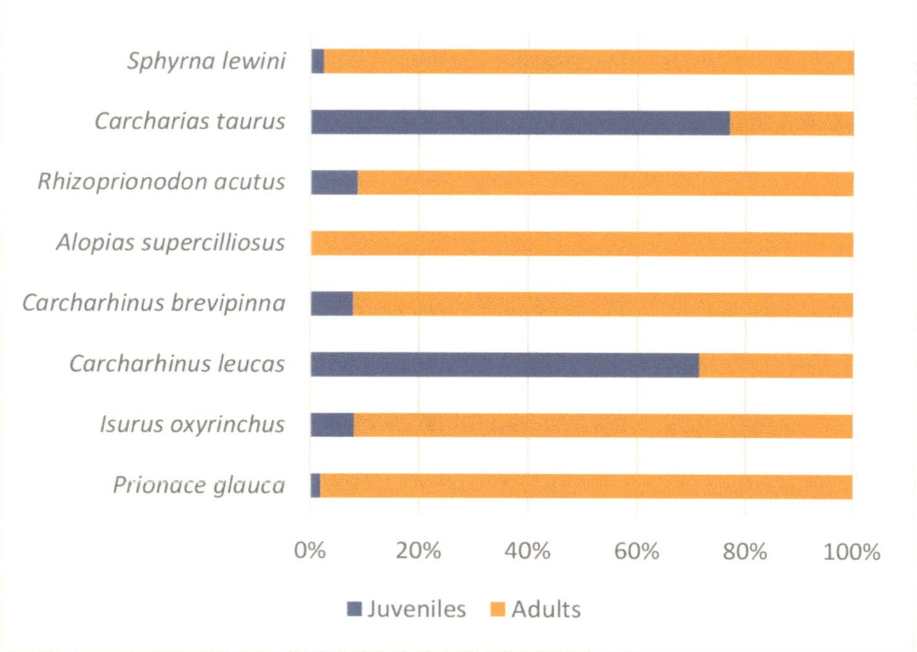

Figure 3.
Composition of adult and juvenile shark species landed during the study period.

Carcharias taurus. Adult individuals dominated the shark species with A. supercilious recording only adult individuals.

Overall, 92% of sharks landed were adults whilst only 8% were in their juvenile stage (**Figure 4**).

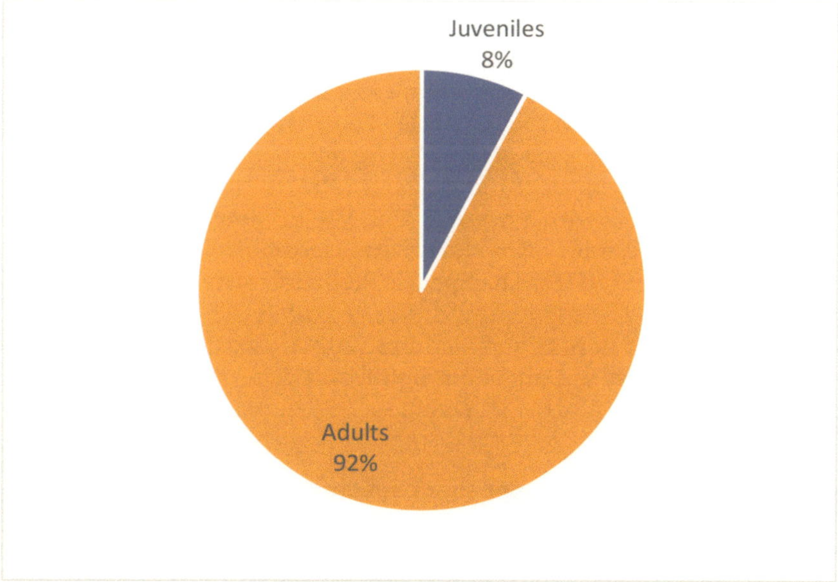

Figure 4.
Maturity composition of shark species landed during the study period.

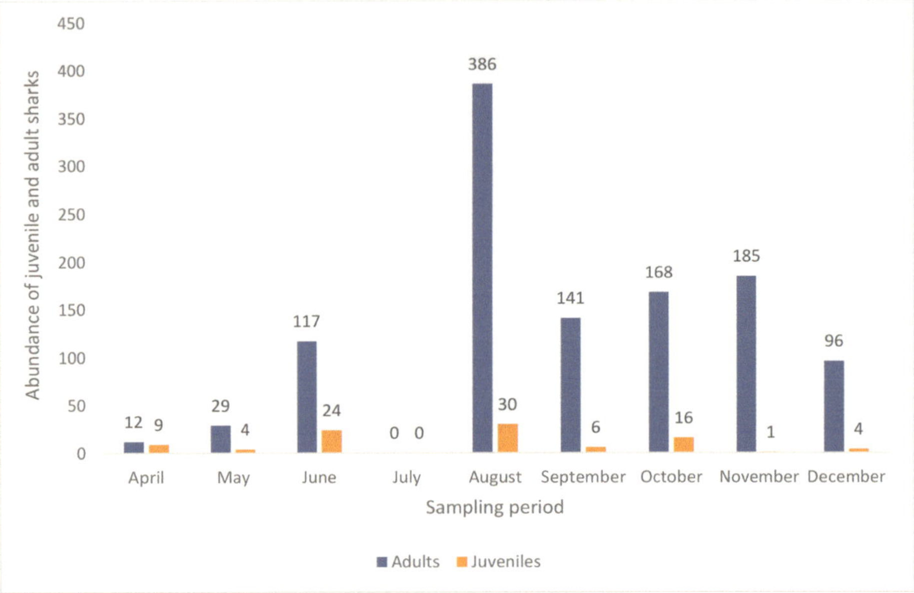

Figure 5.
Abundance of juveniles and adult shark species obtained during the sampling period.

Analysis of the maturity composition of the sharks landed in the studies area shows that August was the month of the most landed shark period with the highest number of adults and juveniles followed by October. A few adults' sharks were landed in April while juvenile sharks are the least landed in November (**Figure 5**).

3.4 Diversity indices

Figure 6 shows the diversity indices for the species obtained during the sampling period. The Species diversity index (D) ranged from 0.3 to 0.7 with a mean of 0.4. The minimum 'D' was recorded in June and December (2021) while the highest 'D' was recorded in October (2021). The Shannon-Wiener index (H) ranged from 0.5 to 1.4 with a mean of 0.9. The minimum 'H' was in June (2021) while the highest 'H' was observed in October (2021). The Species Evenness index (J) ranged from 0.4 to 0.7 with a mean of 0.5. The minimum 'J' was recorded in May, June, August, September, November, and December (2021) while the highest 'J' was recorded in April (2021). The Species Richness index (d) ranged from 0.6 to 1.1 with a mean of 0.9. The minimum 'd' was recorded in June (2021) while the highest 'd' was recorded in May and October (2021). The diversity indices recorded during the study did not show significant difference over sampling period (ONEWAY ANOVA, df = 31, f-value = 1.55, p-value = 0.199).

3.5 Exploitation status of dominant shark species

Figure 7 shows the exploitation status of dominant shark species in terms of length-based measurement. The dominant species from the study was *Prionace glauca* which recorded the highest L_{max}, L_{opt}, L_{inf} and L_m while *Carcharhinus leucas* recorded the lowest L_{max}, L_{opt}, L_{inf} and L_m (**Figure 7**).

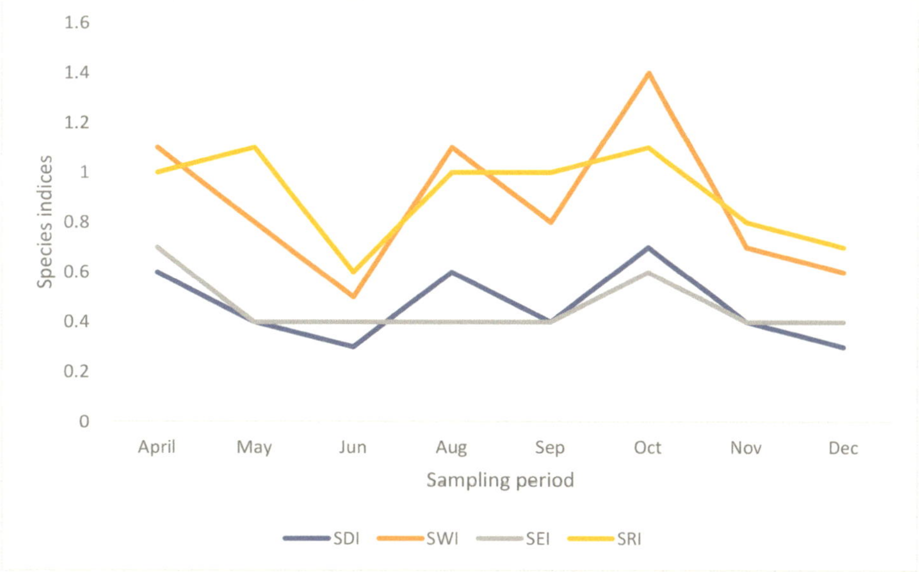

Figure 6.
Diversity indices during the sampling period.

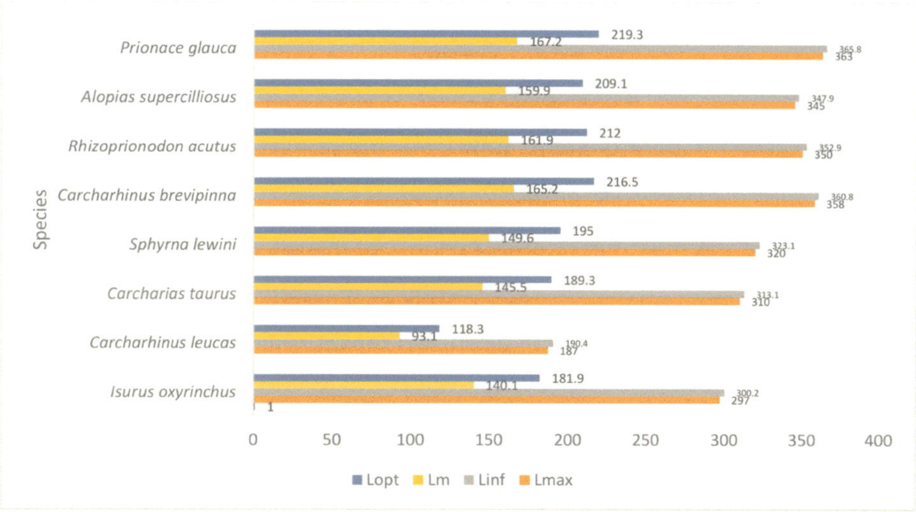

Figure 7.
Parameters for exploitation status of species from the study.

4. Discussion

4.1 Species abundance

From the study conducted, eight (8) shark species were identified at the three sampling locations in the Western region (Shama and Axim) and Greater Accra region (Tema).

The number of species identified in the current study was in variance with other studies. Previous studies by [3, 12, 21] were similar to that reported in other works. Based on the aforementioned studies findings from the current study were similar to that reported in other works. The reasons for the variation in findings in comparison to other studies could be a result of the following factors, environmental factors, time of sampling, sampling duration, depth and type of fishing gear, biological activities of fish species, geographical location, the possibility of tear of fishing gears and the intensity of fishing activities [22]. Concerning the environmental factors, [23] stated that human impact, however, has become a driving force in shaping the spatiotemporal patterns of species abundance and distribution through direct and indirect effects of fishing exploitation, climate change. These impacts have promoted the immigration and expansion of thermophilic taxa, habitat destruction, and pollution which have affected the populations of cartilaginous fish. Multiple species that are now extremely rare or no longer present in the different study areas were a result of prolonged and intense fishing exploitation in the region. Similarly, [24] mentioned that marine fish abundance and distributions are dependent upon a variety of biotic and abiotic factors and may change temporally and/or spatially. According to [25], changes in species assemblages are partly reliant on changes in environmental conditions. For instance, the tide has been suggested as an important factor in influencing the seasonal abundance of fish in the Gulf of Mexico [25]. It is worthy to note that the distribution of sharks responds to seasonal changes, light levels, food availability, predator avoidance, various water quality parameters, and reproductive purposes [24].

Constantly varying the population of the fish assemblages at the shore zone as a result of the changes in the geology of the shore zone is another difference in species distribution variation. Differences in sampling technique, length, and mesh size of gears used are known to affect the abundance of species encountered [26]. Nunoo et al. [25] reported that the duration of sampling influences the abundance of species caught. For instance, [21] over a period of eleven (11) months recorded twenty-three (23) species of sharks from both the artisanal fisheries and trawlers compared to the current studies, which buttress the ascertain that sampling period affects the abundance of species.

In terms of numerical abundance of species the shark (*Prionace glauca*) was the most dominant species within the studied area. Similarly [12] stated that out of the seven (7) species identified, *P. glauca* accounted for the majority of the catch which conforms to the findings from the current study. Depending on the fisheries, areas, and seasons, the blue shark catches can be very significant in the overall catch and some specific cases can account for more than 50% of the total fish catch and around 85–90% of the total elasmobranch catch [27]. The high abundance of blue shark from the current study suggests that the environmental conditions of the marine waters of Ghana are conducive to its sustenance. However, to ensure the sustainability of this dominant shark species in the marine waters of Ghana from future collapse, there is a need for an assessment of its status in the marine waters of Ghana to be conducted. Findings from such studies will help in ensuring proper management measures are drafted and implemented for the sustenance of this species.

4.2 Maturity composition

In Ghana, information on the maturity stages of shark species is lacking, therefore, findings from this study will serve as a baseline for the management of shark fishing in Ghana. According to [28], for many species the percentages of matured specimens in the catch were inversely related to the maximum size. The maturity stage of the sharks recorded from the current study shows that 92% of sharks landed were adults whilst only 8% were in their juvenile stage. The findings were in agreement with a study done by [3] who revealed that 63–76% of the shark species were in the matured stage. However, the finding from the present study was in variance with studies by [21] who reported that about 90% of the species landed were within the juvenile stage. Also, [29] documented that the majority of the shark species captured were juveniles. The reasons for the observed variation in the composition of matured and juvenile shark species landed by artisanal fishermen could be attributed to factors such as seasons, fishing gears and study area. For instance, [29] stated that the highest catches are mostly observed during a reproductive aggregation period from April to June on the Campeche coast, with the adults of both sexes being the most commonly landed specimens.

Consequently, harvesting more adults than juveniles could reduce the recruitment potential of shark species into the stock leading to severe biological and economic repercussions on both the ecosystem and dependent fishing households. This is because elasmobranchs typically have a relatively slow life history due to their large body size, late maturity, slow growth, and low fecundity, which results in low population growth rates. These traits make them exceptionally vulnerable to overfishing and typically result in decreased chances of recovery from population decline and can also lead to stock collapse. The results obtained from the current study could also show that fishers apply destructive fishing methods just to catch more of the adults in order to meet their maximum profits for the fishery, signifying the over-reliance on shark for survival with limited options for alternative livelihoods. Hence, there is a need for the development of proper management measures geared toward a sustainable shark population in the marine waters of Ghana.

The monthly variation for the maturity composition could be alluded to a plethora of factors such as fishermen targeting the matured specimens for economic benefits and the aftermath effect of the upwelling period. Targeting mostly matured individuals by fishermen could reduce the number of species to be recruited into the shark fishery, leading to possible collapse in the future, especially in the absence of proper management measure.

4.3 Diversity indices

Estimation of diversity indices in fishery studies can be useful as changes can be detected in the structure of commercially exploited populations. The Shannon Weiner Index (H) for dominance species range obtained from the current study ranged from 0.5 to 1.4 with a mean of 0.9. The mean of the H for dominance species reported by [24, 30] was 3.81 and 1.1 respectively. The values obtained from the current study were lower than the above-mentioned researchers' values. According to [16], the H value above 3 indicates that the structure of the habitat is stable and balanced and the values below 3 indicate that there are pollution and degradation of habitat structure. From the study, the H was less than 3.0 which indicates possible pollution with the existence of some level of habitat degradation. This could be characterised by the higher number of fishing vessels and thus causing a higher level of fishing and overexploitation.

Species Eveness Index is an important component of diversity indices and expresses the uniform distribution of individuals among different species [31]. The 'J' ranged from 0.4 to 0.7 and with a mean of 0.5 from the current studies. Comparably, the 'J' recorded by [32] in the Mexico waters for Atlantic Sharpnose Sharks was 0.54 as the mean and ranged from 0.34 to 0.69, and that for Bull Sharks was 0.52 as the and ranged from 0.20 to 0.83. The values obtained from the current study was in variance to values reported by the aforementioned researchers. 'J' values close to one (1) indicate an even distribution of a species within an ecosystem, while values closer to zero (0) indicate a site preference [32]. From the current studies, the 'J' was closer to zero (0) indicating site preference for the species. The blue shark was the dominant species (68%) in the study area, which is an indication of site preference for this species.

The Species diversity index (D) refers to its relative importance in its habitat which determines the degree of influence of the species on the ecosystem [22]. The 'D' obtained from the current study ranged from 0.3 to 0.7. and with a mean of 0.4. The mean of the 'D' reported by [33] was 0.04 which was lower than the mean of the 'D' obtained from the current study. Habitats with more interference, tend to have high 'D' (> 0.6), consisting of only one or a few species, and relatively large populations [22]. In this survey, the mean of the 'D' from the current study was 0.4 indicating a relatively stable habitat with less interference hence the composition of dominant species is relatively balanced and not dominated by only one (1) or two (2) particular species. According to [34], Species richness is a diversity of order 0 (which means it is completely insensitive to species abundances); the higher the value, the greater the diversity. The mean of 'd' from the current study (0.9) was closer to one (1) representing an infinite diversity.

The reason for the monthly variation among the various diversity composition could be both the major and minor upwelling season and the aftermath of the close season that was in July and August. Also, migrating pattern of fishes, the difference in sampling period, type of fishing gear could serve as potential factors [22, 35]. Amponsah et al. [22] noted that the variation in species diversity resides in the level of nutrients influx from the coastal lagoon linking each of these marine environments and can be attributed to the linkage of the marine environment and coastal lagoon systems which ensures the exchanges between fresh and marine waters. Tavares and Arocha [30] observed that the highest diversity levels are likely related to the oceanographic factors associated with high marine productivity and the most important factors are the seasonal upwelling.

4.4 Exploitation status

Research on the exploitation status of sharks has not been done or published in Ghana. Therefore, this current study is the first research on the exploitation status of sharks in Ghana waters. The exploitation status from the current status was based on three indicators to deal with overfishing proposed by [36]. Indicator one (1) was described as letting them spawn and it was measured as a percentage of mature specimens in the catch. The target would be to let all fish spawn at least once before they are caught to rebuild and maintain healthy spawning stocks. Comparing this analysis to the current study from **Table 2**, the majority of the species were matured except for *Carcharhinus leucas*. However, this analysis seems to favour the current study indicating a healthy spawning stock. Indicator two (2) is the 'Let them grow!' and is measured as the percentage of fish caught at the optimum length. This is where the Optimum length is typically a bit larger than the length at first maturity. From the study, the optimum length (L_{opt}) estimated for all the species was larger than the length at first, maturity (L_m) which satisfies indicator 2. Indicator three (3) is described as allowing the mega spawners to live!' and is measured as the percentage of old, large fish in the catch.

Species	Mega spawners (%)	Percentage of juveniles (%)	Matured Specimen (%)
Isurus oxyrinchus	78.0	16.7	83.3
Carcharhinus leucas	66.6	61.5	38.5
Carcharias taurus	23.0	22.2	77.8
Sphyrna lewini	61.9	37.5	62.5
Carcharhinus brevipinna	42.5	37.7	62.3
Rhizoprionodon acutus	28.0	37.6	62.4
Alopias supercilliosus	100	0.0	100
Prionace glauca	38.4	29.7	70.3

Table 2.
Percentage of mega spawners, juveniles, and matured specimen.

From the current study, the percentage of matured specimen caught within the study period was 92%, and the juveniles as only 8%, which is highly greater than the threshold of 20% of the catch which indicates a healthy size structure. According to [36] if L_m is less than L_{opt} less than L_{inf} ($L_m < L_{opt} < L_{inf}$), it conveys that, all fish are given a chance to reach the size of maximum growth rate (and reproduce before being caught, so growth and recruitment overfishing are theoretically impossible and impact on expected life-time fecundity per recruit is reduced. The results from the current study fit perfectly into the framework as shown in **Table 2** which buttresses the fact that the stock of these species is in healthy condition. In effect, this condition improves the ecosystem resilience and stability which serves as preconditions for reliable ecosystem services.

5. Conclusion

The study aimed to assess the abundance and distribution of sharks in Ghana's coastal area. Overall, eight (8) species of sharks were obtained with *Prionace glauca* (68%), as the most dominant species with both *Carcharhinus leucas* (1%) and *Alopias supercilious* (1%) as the fewer dominate species. The percentage of the maturity composition of adult to juvenile sharks is 92:8% with August happens to be the month of the most landed shark period with the highest number of adults and juveniles landed possibly as a result of the close season in July, followed by October. The diversity indices showed the marine environment is moderately polluted, the species structure is evenly distributed with infinite diversity and not dominated by only one (1) species. The exploitation status of the species based on length measurement indicates that the stock of the shark species is in healthy condition.

Acknowledgements

The authors express sincere gratitude to the fishermen and the personnel from Axim, Dixcove, and Tema fisheries commission for their assistance provided during the measurement and identification of the samples.

Conflicts of interest

No conflicts of interest.

Author details

Samuel K.K. Amponsah*, Rachael Ackah, William Dzisenu Amekor, Asiedu Berchie and Andrews Apraku
Department of Fisheries and Water Resources, University of Energy and Natural Resources, Sunyani, Ghana

*Address all correspondence to: samuel.amponsah@uenr.edu.gh

References

[1] Abdulla A. Status and Conservation of Sharks in the Mediterranean Sea. 2014. Available from: http://www.redlist.org/info/categories_criteria.html

[2] Roff G, Doropoulos C, Rogers A, Bozec YM, Krueck NC, Aurellado E, et al. The ecological role of sharks on coral reefs. Trends in Ecology and Evolution. 2016;**31**(5):395-407. DOI: 10.1016/j.tree.2016.02.014

[3] Seidu I, Brobbey LK, Danquah E, Oppong SK, van Beuningen D, Seidu M, Dulvy NK. Fishing for survival: Importance of shark fisheries for the livelihoods of coastal communities in Western Ghana. Fisheries Research. 2022;**246**:106-157. DOI: 10.1016/j.fishres.2021.106157

[4] Blanco-Parra M-D-P. Line: Ecological effects of fishing on socio-environmental ecosystems (Project: functional structure and trophic ecology of fish) View project. 2008. Available from: https://www.researchgate.net/publication/299077044

[5] Camhi M, Pikitch EK, Babcock EA. Sharks of the Open Ocean: Biology, Fisheries and Conservation. USA: Blackwell Science; 2008. pp. 166-192

[6] Baum JK, Myers RA, Kehler DG, Worm B, Harley SJ, Doherty PA. Collapse and conservation of shark populations in the Northwest Atlantic. Science. 2003;**299**(5605):389-339

[7] Lam VYY, Sadovy De Mitcheson Y. The sharks of South East Asia - Unknown, unmonitored, and unmanaged. Fish and Fisheries. 2011;**12**(1):51-74. DOI: 10.1111/j.1467-2979.2010.00383.x

[8] Barker MJ, Schluessel V. Managing global shark fisheries: Suggestions for prioritizing management strategies. Aquatic Conservation: Marine and Freshwater Ecosystems. 2005;**15**(4):325-347. DOI: 10.1002/aqc.660

[9] Martins APB, Feitosa LM, Lessa RP, Almeida ZS, Heupel M, Silva WM, et al. Analysis of the supply chain and conservation status of sharks (Elasmobranchii: Superorder Selachimorpha) based on fisher knowledge. PLoS One. 2018;**13**(3):1-15. DOI: 10.1371/journal.pone.0193969

[10] Van Waerebeek K, Ofori-Danson PK. A first checklist of cetaceans of Ghana, Gulf of Guinea, and a shore-based survey of interactions with coastal fisheries. In: Paper SC/51/SM35 presented to the 51st Annual Meeting of the International Whaling Commission Scientific Committee, Grenada. 1999

[11] Sall A, Failler P, Drakeford B, March A. Fisher migrations: Social and economic perspectives on the emerging shark fishery in West Africa. African Identities. 2021;**19**(3):284-303. DOI: 10.1080/14725843.2021.1937051

[12] Agyeman NA, Blanco-Fernandez C, Steinhaussen SL, Garcia-Vazquez E, Machado-Schiaffino G. Illegal, unreported, and unregulated fisheries threatening shark conservation in African waters are revealed by high levels of shark mislabelling in Ghana. Genes. 2021;**12**(7):1002. DOI: 10.3390/genes12071002

[13] Schneider W. FAO Species Identification Sheets for Fishery Purposes. Field Guide to the Commercial Marine Resources of the Gulf of Guinea. Prepared and Published with the

Support of the FAO Regional Office for Africa. Rome: FAO; 1990. p. 268

[14] Oliveira-Santos LGR, Graipel ME, Tortato MA, Zucco CA, Cáceres NC, Goulart FV. Abundance changes and activity flexibility of the oncilla, *Leopardus tigrinus* (Carnivora: Felidae), appear to reflect avoidance of conflict. Zoologia (Curitiba). 2012;**29**:115-120

[15] Ramos S, Cowen RK, Re P, Bordalo AA. Temporal and spatial distribution of larval fish assemblages in the Lima estuary (Portugal). Estuarine, Coastal and Shelf Science. 2006;**66**: 30314

[16] Shannon CE, Weaver W. The Mathematical Theory of Communication. Urbana: University of Illinois Press; 1949

[17] Margalef R. Some concepts relative to the organization of plankton. In: Barnes HB, editor. Oceanography and Marine Biology: An Annual Review. Vol. 5. London: George Allen & Unwin; 1967. pp. 257-289

[18] Ogbeibu AE. Biostatistics: A Practical Approach to Research and Data Handling. Benin City: Mindex Publishing Company Limited; 2005

[19] Pielou EC. An Introduction to Mathematical Ecology. New York, USA: Wiley-Inter-science; 1969

[20] Froese R, Binohlan C. Empirical relationships to estimate asymptotic length, length at first maturity and length at maximum yield per recruit in fishes, with a simple method to evaluate length frequency data. Journal of Fish Biology. 2000;**56**(4):758-773

[21] Kiilu BK, Kaunda-Arara B, Oddenyo RM, Thoya P, Njiru JM. Spatial distribution, seasonal abundance and exploitation status of shark species in Kenyan coastal waters. African Journal of Marine Science. 2019;**41**(2):191-201

[22] Amponsah SKK, Ameyaw A, Asare C. Agriculture, livestock and fisheries multivariate analysis of abundance and distribution of fish species in coast of Ghana, West Africa. Review article. Research in Agriculture, Livestock and Fisheries. 2022;**9**(1): 57-69. Available from: www.agroaid-bd.org/ralf

[23] Serena F, Abella AJ, Bargnesi F, Barone M, Colloca F, Ferretti F, et al. Species diversity, taxonomy and distribution of Chondrichthyes in the Mediterranean and Black Sea. The European Zoological Journal. 2020; **87**(1):497-536

[24] Parsons GR, Hoffmayer ER. Seasonal changes in the distribution and relative abundance of the *Atlantic Sharpnose* Shark *Rhizoprionodon terraenovae* in the north central Gulf of Mexico. Copeia. 2005a;**4**:914-920. DOI: 10.1643/0045-8511(2005)005[0914:SCITDA]2.0.CO;2

[25] Nunoo FKE, Eggleston DB, Vanderpuye CJ. Abundance, biomass, and species composition of nearshore fish assemblages in Ghana, West Africa. African Journal of Marine Science. 2006; **28**(3–4):689-696. DOI: 10.2989/18142320609504217

[26] Lasiak T. Structural aspects of the surf-zone fish assemblage at King's Beach, Algoa Bay, South Africa: Short-term fluctuations. Estuarine, Coastal and Shelf Science. 1984:**18**(4):459-483. DOI: 10.1016/0272-7714(84)90084-2

[27] Coelho R, Mejuto J, Domingo A, Yokawa K, Liu KM, Cortés E, et al. Distribution patterns and population

structure of the blue shark (Prionace glauca) in the Atlantic and Indian Oceans. Fish and Fisheries. 2018;**19**(1): 90-106. DOI: 10.1111/faf.12238

[28] Harry AV, Macbeth WG, Gutteridge AN, Simpfendorfer CA. The life histories of endangered hammerhead sharks (Carcharhiniformes, Sphyrnidae) from the east coast of Australia. Journal of Fish Biology. 2011;**78**(7):2026-2051

[29] Pérez-Jiménez JC, Wakida-Kusunoki A, Hernández-Lazo C, Mendoza-Carranza M. Shark-catch composition and seasonality in the data-poor small-scale fisheries of the southern Gulf of Mexico. Marine and Freshwater Research. 2020; **71**(9):1182-1193. DOI: 10.1071/ MF19184

[30] Tavares R, Arocha F. Species diversity, relative abundance, and length structure of oceanic sharks caught by the Venezuelan longline fishery in the Caribbean Sea and western-central Atlantic. Zootecnia Trop. 2008;**26**(4): 489-503

[31] Bibi F, Ali Z. Measurement of diversity indices of avian communities at Taunsa Barrage Wildlife Sanctuary, Pakistan. The Journal of Animal & Plant Sciences. 2013;**23**(2):469-474

[32] Altobelli AN, Szedlmayer ST. Migration and residency of sandbar, Atlantic sharpnose, bull, and nurse sharks in the Northern Gulf of Mexico. North American Journal of Fisheries Management. 2020;**40**(5):1324-1343. DOI: 10.1002/nafm.10501

[33] Doane MP, Haggerty JM, Kacev D, Papudeshi B, Dinsdale EA. The skin microbiome of the common thresher shark (*Alopias vulpinus*) has low taxonomic and gene function β-diversity. Environmental Microbiology Reports. 2017;**9**(4):357-373. DOI: 10.1111/1758-2229.12537

[34] Gotelli NJ, Chao A. Measuring and estimating species richness, species diversity, and biotic similarity from sampling data. In: Encyclopedia of Biodiversity. Vol. 5. Second ed. Waltham, MA: Elsevier Inc. Academic Press; 2013. pp. 195- 211. DOI: 10.1016/ B978-0- 12-384719-5.00424-X

[35] Henderson AC, McIlwain JL, Al-Oufi HS, Al-Sheili S. The sultanate of Oman shark fishery: Species composition, seasonality, and diversity. Fisheries Research. 2007;**86**(2–3):159-168. DOI: 10.1016/j.fishres.2007.05.012

[36] Froese R. Keep it simple: three indicators to deal with overfishing. Fish and Fisheries. 2004;**5**:86-91

Chapter 4

Elasmobranches and Chimaeras in Syria: Past, Present, and Future

Adib Saad and Hasan Alkusairy

Abstract

In this chapter, we review and discuss the cartilaginous species richness in Syrian offshore waters (Eastern Mediterranean coast) through a careful review of published taxonomic studies, historical data on species occurrence, and analysis of scientific surveys carried out over the last 10 years. The revised species produced for the Syrian coast, in this chapter, is debated in the context of current taxonomic disputes and the occurrence of species. Annual catches of each species over two consecutive years are also presented. Threats were estimated based on field surveys and observations conducted during the last three decades, particularly the results of the qualitative composition of the catch and the maturity status of individuals that were carried out during the years 2015–2018; extensively classified conservation measures have also been proposed. We listed a total of 49 species, representing 25 families and 34 genera that are currently present in Syrian marine waters. This number includes 23 species of sharks, 24 batoids (after we considered the species *Pristis pectinata* an extinct species from the Syrian marine water) and two chimaeras. In addition, the review serves as a reference for future conservation assessments of cartilaginous fishes in the region and a guide for decision-makers when promoting sustainable exploitation of fisheries resources within an ecosystem-based framework.

Keywords: sharks, batoids, chimaeras, Mediterranean sea, Syria

1. Introduction

Cartilaginous fishes, also called chondrichthyans, are characterized by a skeleton composed of cartilage. In addition, their skin is covered with placoid scales, which look a lot like a vertebrate tooth. This is why these animals have a rough texture to the touch. Their mouth contains many partially calcified teeth. These are not fused to the jaw, so they also have several spares. These fishes have two nostrils, a blowhole, and 5–7 gill openings. Their eyes do not have eyelids, but some of them have a light, transparent membrane, called the nictitating membrane, which performs the same function. Another main characteristic is the presence of pelvic fins in the lower part of their body, which serves as the reproductive organs of these fish. These fins, also called claspers, allow them to deposit gametes (sperm) inside the female. That is why only males have them. Cartilaginous fish are divided into three groups: A. Sharks: Most sharks are characterized by a streamlined spindle-shaped body and a well-developed caudal fin composed of two lobes, with the upper lobe longer than the lower lobe. They have five- to seven-gill openings behind the head on each side of the body. B. Batoids (Skates, rays, sawfishes, guitarfishes): Skates and rays are characterized by their flattened shape dorsally and ventrally, as most species live sedentary on the bottom. There are five- or six-gill slits on the ventral side of the body. C. Chimaeras: They have strange shapes, and unlike other cartilaginous fish, mythical fish have one pair of gill openings covered by a skin fold.

From a taxonomic point of view, the class of cartilaginous fishes is divided into two subclasses: the subclass Elasmobranchs (sharks and batoids (rays and skate)) and the subclass Holocephali. To date, 1226 species of elasmobranches have been described (537 species of sharks belonging to 34 families, 689 species of batoids belonging to 20 families) and 56 species of Halocephali belonging to three families [1–3]. Cartilaginous fish are widely distributed in marine ecosystems, and sometimes they are also found in freshwater environments [4]. Syrian marine waters form the north-eastern part of the Levant basin in the eastern part of the Mediterranean Sea, as this basin is characterized by high salinity (39.5‰) and temperature (26°C) [5], the arrival of new species through the Suez Canal, especially of Indo-Pacific origin, has increased species richness in the eastern basin [6].

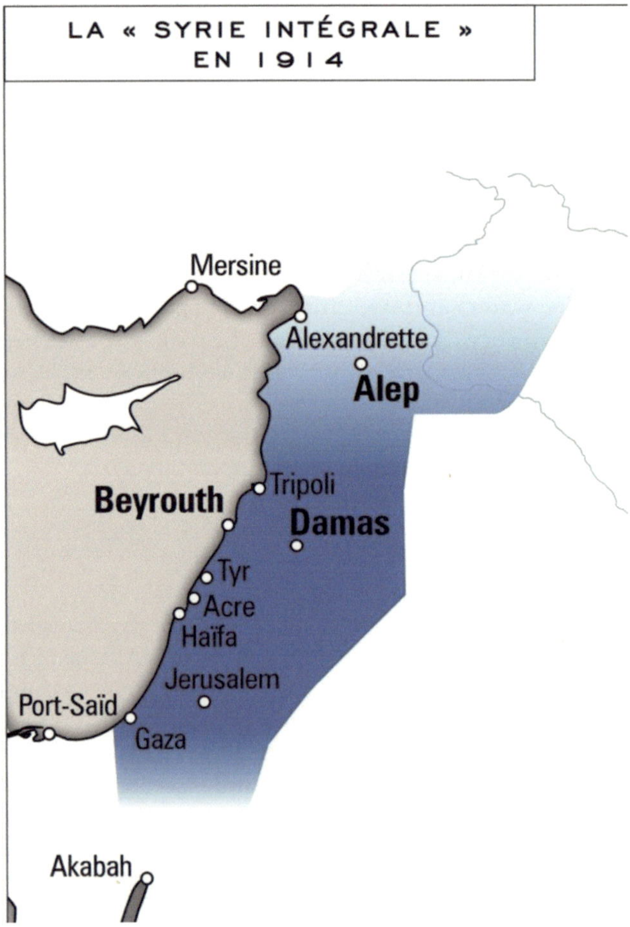

Figure 1.
The integral Syrian coast where grovel made a survey of sharks and other marine organisms during the period 1929–1931 [7].

Before 1920, Syria's natural coast constituted the eastern coast of the Mediterranean, extending from the city of El-Arish in Egypt to the city of Iskenderun in Turkey (**Figure 1**), while currently this coast is limited to 183 km in length from the Lebanese border in the south to the Turkish border in the north [7].

The general marine fauna of the coasts of Syria can be considered as formed by the fauna of the Mediterranean, which is superimposed, in a way, an important part of the fauna of the Red Sea and the Indian Ocean. A large number of species from these last seas have found favorable biological conditions on the Syrian (including Palestinian and Lebanese) coasts (**Figure 1**). The continuous monitoring studies during the last decades confirm that the number of these forms coming through the Suez Canal grow and develop permanently over time because one or more species of Indo-Pacific origin are recorded each week in the Levantine basin in general and on the Syrian coast in particular. In addition, many species of fish of Atlantic origin are recorded in the eastern Mediterranean and the Syrian coasts. These new introductions and invasions of exotic species (whether from the Red Sea or the Atlantic Ocean) are mainly due to climate change and human activities [8, 9].

In this chapter, we revise and discuss the chondrichthyan species richness and their records. Through an accurate review of published taxonomic studies, historical data on species occurrence and analyses of scientific survey produce a revised list of species whose presence in the Syrian coast (**Figure 2**) is confirmed or highly probable and discussed on current taxonomic and occurrence disputes on the species that are instead rarer or claimed to be locally extinct.

The review serves as a reference for future conservation assessments of cartilaginous fishes in Syrian marine waters and a guide for decision-makers when promoting sustainable exploitation of fisheries resources within an ecosystem-based framework, explaining the reasons for their prolonged absence in reports.

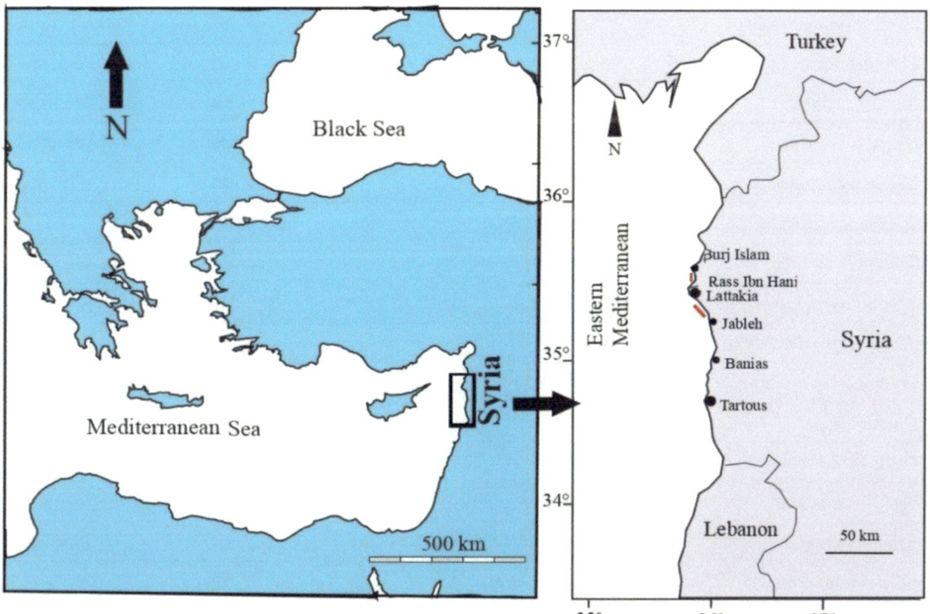

Figure 2.
Map of the Mediterranean showing the study area, Syrian coast, on which the most important cities and ports are located in which the landing of fish catch every day (= two nursery areas for several species of sharks).

2. Materials and methods

The information was collected from two main sources: 1- the reference study of previously published scientific literature and 2- field studies carried out by the authors during the last two decades. Our field studies have investigated the species diversity, abundance, longevity, sex ratio, and nursery potential of caught sharks. Samples were obtained by all fishing methods (longlines, nets, and nets) during the period 2001–2003 and 2016–2021 in Syrian waters. For scientific nomenclature, we followed the online database of the Eschmeyer's Catalog of Fish [10], which is the authoritative reference for taxonomic fish names, together with the world record of marine species (WoRMS) and Fishbase [11, 12]. Regarding the study of catch composition and species abundance level, species abundance (abundance index) was determined according to the numbers caught during the study as follows: dominant: >1000 individuals, common: 200–1000 individuals, frequent: 100–199 individuals, occasional: 10–99 and rare individuals: less than 10 individuals.

During the 3 years (2011, 2015, and 2016), daily surveys were made on species number, weight, and size structure caught by all fishing gears in the main landing of chondrichthyans on Syrian coast, the ecological indices; diversity index D, qualitative richness factor QR, and evenness factor E were used to evaluate the levels of shark diversity within the study region by all gear types [13, 14].

Species	Size at first maturity in the Syrian marine waters (mm)	Refs. of biological studies	Percentage of juvenile individuals	The main fishing method
Hexanchus griseus			**99**	long line
Heptranchias perlo	f: 970, m: 810	[15]	**63**	Trawl + long line
Carcharhinus plumbeus			**87**	Trawl + long line
Carcharhinus obscurs			**85**	long line
Isurus oxyrinchus			**100**	Fishing nets
Mustelus mustelus	f: 105, m: 101	[16]	**86**	Fishing nets
Galeus melastomus			**16**	Trawl
Syliorhinus canicula	f: 380, m: 370	[15]	**51**	Trawl
Squalus blainvillei			**55**	Trawl
Squalus megalops			**60**	Trawl
Alopias superciliosus			**100**	Long line
Centrophorus uyato	f: 867, m: 792	[15]	**84**	Trawl + long line
Dalatias licha			**38**	Trawl
Oxynotus centrina			**100**	Trawl
Squatina aculeata			**46**	long line
Squatina squatina			—	long line
Glaucostegus cemiculus	f: 870, m: 860	[16]	**52**	Trawl + long line
Rhinobatus rhinobatos			**24**	Trawl + long line
Dipturus oxyrinchus	f: 795, m: 709	[17]	**68**	Trawl
Raja clavata	f: 575, m: 480	[18]	**80**	Trawl

Species	Size at first maturity in the Syrian marine waters (mm)	Refs. of biological studies	Percentage of juvenile individuals	The main fishing method
Raja radula			—	Trawl
Raja miraletus			—	Trawl
Leucoraja circularis			—	Trawl
Leucoraja fullonica			—	Trawl
Gymnura altavela	*F: 961, m: 771*	[19]	**50**	Trawl + long line
Aetomylaeus bovinus			**50**	Trawl + long line
Myliobatis aquila			—	Trawl
Rhinoptera marginata			**91**	Trawl
Mobula mobular			**100**	long line
Dasyatis pastinaca			**54**	Trawl + long line
Dasyatis tortonesei			**81**	Trawl + long line
Bathytoshia centoura			—	Trawl + long line
Pteroplatytrygon violacea			**80**	Trawl
Himantura uarnak			—	Trawl + long line
Taeniura grabat			—	Trawl + long line
Tetronarce. nobiliana			**100**	Trawl
Torpedo marmorata			**100**	Trawl
Torpedo sinuspersici			—	Trawl
Chimaera monstora			—	Trawl

Table 1.
The status of cartilaginous species in our field study that was carried out during the years 2015–2016; size at first maturity (mm), percentage of juvenile individuals and main catch method in the Syrian marine waters (f: Females, m: Males, TL: Total length, DW: Disc width).

Biodiversity indicators of caught cartilaginous fish:

- Diversity index (D): $D = -\sum Pi \times Ln(Pi)$.
- Pi: the ratio of the number of individuals of a species to the total number of individuals, Ln(Pi): the natural logarithm of the ratio of the number of individuals of a species.
- Qualitative richness (QR): $QR = (S-1)/Ln(N)$.
- S: the total number of species, N: the number of individuals in the sample.
- Evenness (E): $E = QR/Ln(S)$.
- QR: coefficient of species richness, S: total number of species, Ln(S): natural logarithm of the number of species.

Juvenile individuals of the species were defined by information on size at first maturity of the Syrian marine fishes or the nearest region to Syrian coast were extracted from peer-reviewed published sources, including 'gray' literature [13, 14]. The reproductive cycle and maturity of eight species of cartilaginous fish were studied in detail during the aforementioned period, and the results were published in scientific journals, which are listed in **Table 1**.

Possible nursery places were defined depending on individual's size of species, density of individuals, and repeat individual's appearance in the fishing area [13, 14].

3. Results and discussion

3.1 Recorded species and diversity

The study of the distribution and taxonomy of sharks in Syrian marine waters were initiated by Cruvel during the period 1927–1930 [20]. This study reported 14 species of cartilaginous fish, including six sharks, seven batoids, and one chimaera. This was followed by a short study on the assessment of demersal fish stocks during the year 1976 [21] during which it confirmed the presence of five species in addition to the previously recorded species by Gruvel [20], and then a study of the qualitative composition of cartilaginous fishes during the period 2002–2004 [22–24], and finally an extensive study on the qualitative and quantitative composition of cartilaginous fishes in Syrian marine waters [13, 14].

The number of species recorded on the Syrian coast is 49 cartilaginous species (23 sharks, 24 batoids, and two chimaera) (**Table 2**). Among them, one species of batoids (*Pristis pectinata* Latham 1794) has disappeared from the Syrian marine waters and perhaps from all the waters of the eastern part of the Mediterranean, as the last record of it in the Syrian waters is 1976 [20, 21]; this number represented about 55.1% of the cartilaginous fish in the Mediterranean Sea (what constitutes 46.9% of all shark species, 63.2% of batoids species, and 100% of chimaeras) (**Table 3**). The cartilaginous fish recorded in the Mediterranean are 89 species composed of 49 sharks, 38 batoids, and 2 chimaeras [33, 34].

Area	Taxa	Order	Family	Genus	Species	%
Mediterranean and Black Sea	Sharks	5	18	27	49	
	Batoids	4	11	19	38	
	Chimaeras	1	1	2	2	
Total		10	30	48	89	
Syrian marine waters	Sharks	5	15	17	23	46.9
	Batoids	4	9	17	24	63.2
	Chimaeras	1	1	2	2	100
Total		10	25	36	49	55.1

Table 2.
Synoptic table showing the number of species belonging to each systematic group of chondrichthyans. The relationship between Syrian marine waters chondrichthyans and Mediterranean is compared in terms of percentage.

Classes and orders	Family and species	Status	Frequency	First record	First reference
Elasmobranchii (cohort Selachii)—Sharks					
Order					
Hexanchiformes					
	Hexanchidae				
	Heptranchias perlo (Bonnaterre, 1788)	N	F	2002	[23]
	Hexanchus griseus (Bonnaterre, 1788)	N	C	2002	[23]
Lamniformes					
	Lamnidae				
	Isurus oxyrinchus Rafinesque, 1810	N	F	2003	[23]
	Cetorhinidae				
	Cetorhinus maximus (Gunnerus, 1765)	N	VR	2011	[25]
	Alopiidae				
	Alopias superciliosus Lowe, 1841	N	F	2003	[23]
Carcharhiniformes					
	Scyliorhinidae				
	Galeus melastomus Rafinesque, 1810	N	F	2003	[22]
	Scyliorhinus canicula (Linnaeus, 1758)	N	F	1929	[20]
	Scyliorhinus stellaris (Linnaeus, 1758)	N	R EN	1929	[20]
	Triakidae				
	Mustelus mustelus (Linnaeus, 1758)	N	R EN	1929	[20]
	Carcharhinidae				
	Carcharhinus obscurus (Lesueur, 1818)	N	F	2003	[23]
	Carcharhinus plumbeus (Nardo, 1827	N	C	2003	[23]
	Sphyrnidae				
	Sphyrna zygaena (Linnaeus, 1758)	N	VR	1929	[20]
Squaliformes					
	Dalatiidae				
	Dalatias licha (Bonnaterre, 1788)	N	F	2002	[23]

Classes and orders	Family and species	Status	Frequency	First record	First reference
	Somniosidae				
	Somniosus rostratus (Risso, 1827)	N	F	2002	[23]
	Etmopteridae				
	Etmopterus spinax (Linnaeus, 1758)	N	R	2016	[26]
	Oxynotidae				
	Oxynotus centrina (Linnaeus, 1758)	N	F	2003	[23]
	Centrophoridae				
	Centrophorus uyato (Rafinesque, 1810) = *C. granulosus* (Bl. et Sch, 1801)	N	F	2003	[20]
	Squalidae				
	Squalus blainville (Risso, 1827)	N	F	2002]	[23]
	Squalus megalops (Macleay, 1881)	REX	R	2002	[23]
	Squalus acanthias (Linnaeus, 1758)	N	VR (EN)	1929	[20]
Squatiniformes					
	Squatinidae				
	Squatina aculeata Cuvier, 1829	N	VR		[23]
	Squatina oculata Bonaparte, 1840	N	R EN	2003	[23]
	Squatina squatina (Linnaeus, 1758)	N	R EN	1929	[20]
(Cohort Batoidea) -Batoids					
Torpediniformes					
	Torpedinidae				
	Torpedo marmorata Risso, 1810	N	R	1929	[20]
	Tetronarce nobiliana (Bonaparte, 1835)	N	R	2003	[23]
	Torpedo sinuspersici Olfers, 1831	AL	R	2003	[23]
	Torpedo torpedo (Linnaeus, 1758)	N	VR EN	1929	[20]
Rhinopristiformes					
	Rhinobatidae				
	Rhinobatos rhinobatos (Linnaeus, 1758)	N	F	1976	[21]

Classes and orders	Family and species	Status	Frequency	First record	First reference
	Glaucostegidae				
	Glaucostegus cemiculus (Geoffroy Saint- Hilaire, 1817)	N	R	2003	[23]
Rajiformes					
	Rajidae				
	Dipturus oxyrinchus (Linnaeus, 1758)	N	F	1929	[23]
	Raja miraletus Linnaeus, 1758	N	F	1929	[20]
	Raja clavata Linnaeus, 1758	N	R	1929	[10]
	Raja radula Delaroche, 1809	N	R	2003	[23]
	Leucoraja circularis (Couch, 1838)	N	V R	2018	[27]
	Leucoraja fullonica (Linnaeus, 1758)	N	VR		[28]
Myliobatiformes					
	Dasyatidae			1928	[20]
	Bathytoshia centroura (Mitchill, 1815)	N	F	2015	[13]
	Dasyatis pastinaca (Linaeus, 1758)	N	F	1928	[20]
	Dasyatis sp. cf. *tortonesei* Capapé, 1975	N	R	2003	[23]
	Himantura leoparda		VR	2021	[29]
	Himantura uarnak (Gmelin 1789)	Al	R	2009	[30]
	Pteroplatytrygon violacea (Bonaparte, 1832)	N	F		[23]
	Taeniurops grabata (Geoffroy Saint-Hilaire, 1817)	N	F	2012	[31]
	Gymnuridae				
	Gymnura altavela (Linnaeus, 1758)	N	F	2003	[23]
	Myliobatidae				
	Myliobatis aquila (Linnaeus, 1758)	N	R	1929	[20]
	Aetomylaeus bovinus (Geoffroy Saint-Hilaire, 1817)	N	R	2003	[23]
	Rhinopteridae				
	Rhinoptera marginata (Geof. St. Hilaire, 1817)	N	R		[10]
	Mobulidae				

Classes and orders	Family and species	Status	Frequency	First record	First reference
	Mobula mobular (Bonnaterre, 1788)	N	R		[23]
Class Holocephali					
Chimaeriformes					
	Chimaeridae				
	Chimaera monstrosa Linnaeus, 1758	N	F	1929	[20]
	Hydrolagus mirabilis (Collett, 1904)	REX		2013	[32]

Table 3.
List of recorded species of cartilaginous fish in Syrian marine water. The status, frequency, the first record, and first publication (N = native; AL = alien; REX = range expanding; C = common; F = frequent; R = rare; VR = very rare; EN = endangered.)

Only two studies on biodiversity evidence for cartilaginous fish were carried out during the years 2015–2016 in Syrian marine waters [13, 14]. In these studies, the ecological indices, diversity index D, qualitative richness factor QR, and evenness factor E were used to evaluate the levels of shark diversity within the study region by all fishing methods (**Table 4**). The means of D, QR, and E through the period of study were (1.73 ± 0.21), (2.03 ± 0.40), and (0.67 ± 0.09), respectively. The seasonal variations of the three ecological indices show that, the highest mean of D values was in summer (2.41 ± 0.09), followed by spring (2.40 ± 0.05), autumn (2.37 ± 0.21), and winter (2.15 ± 0.13). Additionally, the highest mean of QR values was in spring (3.95 ± 0.13), followed by autumn (3.22 ± 0.73), summer (3.13 ± 0.44), and winter (3.13 ± 0.11), while the highest mean of E values was in summer (0.75 ± 0.03), followed by autumn (0.75 ± 0.06), spring (0.69 ± 0.03), and winter (0.68 ± 0.05). The highest values of D and QR in summer and spring are because of caught high numbers of species S and numbers of species individuals N in these seasons, which can be explained by the movements of species individuals toward coast to bear and search for feeds, while the highest value of E was in winter due to the low numbers of species S. Estimates of diversity levels in fishery studies can be useful as changes can be detected in the structure of commercially exploited populations. Diversity indices are rarely applied in chondrichthyans fisheries.

Index/factor	Sharks catch (8456 individuals, 17 species)	Batoids catch (14,724 individuals, 19 species)	Total catch (23,197 individuals, 37 species)
Diversity index D	1.73 ± 0.21	1.82 ± 0.16	2.33 ± 0.16
Qualitative richness factor QR	2.03 ± 0.04	1.67 ± 0.24	3.36 ± 0.46
Evenness factor	0.67 ± 0.09	0.72 ± 0.05	0.72 ± 0.05

Table 4.
Average (mean ± SD) of diversity index D, qualitative richness factor QR and evenness factor E values for a total catch of cartilaginous fish, sharks catch and batoids catch during 2015–2016 [13, 14].

3.2 Cartilaginous fish landings in 2015 and 2016

The average of total catch amounted during 2 years to 61,912 kg. The largest amount of catch was in the month of September (**Figure 3**), as it reached in the first year 18,182 kg and in the second year 9645 kg with an average of 13,914 kg/year, followed by October with an amount of 6413 kg in the first year and in the second year 8263 kg with an average of 8138 kg, while the lowest amount of catch was in the first and second years in January 2079 kg and 2410 kg, respectively, with an average of 2244 kg. The species *Hexanchus griseus* ranked first in terms of weight in the catch, as its quantity in the first and second years amounted to 15,963 kg and 17,129 kg, with an average of 16,549 kg, followed by the species *Carcharhinus plumbeus* with 14,445 kg and 7886 kg, respectively, in the first and the second year. The average was 11,166 kg. The total catch of the two species of sharks, *H. griseus* and *C. plumbeus*, constituted about 40% of the total average catch in both years (2015 and 2016). The percentage of cartilaginous fish landing during 3 years 2011 [22] and 2015–2016 [13] amounted to 3.68% of the amount of marine fish caught in Syrian marine waters. The species *Centrophorus uyato* (16.8%) occupied the highest percentage of cartilaginous fish catches (in terms of number), followed by *Rhinobatus cemiculus* (10.8%) and *Squalus* sp. (9.3%). In terms of weight, *H. griseus* occupied the highest percentage (30.6%), followed by *C. plumbeus* (25.2%), then *Rhinobatus cemiculus* (8.2%) [13].

As for the average total number of individuals, it reached 11,373 individuals. The largest number of individuals of the species in the catch in the first and second years was in May with an average of 1947 individuals (**Figure 4**). This was followed by April, with an average of 1578 individuals. As for the last month in terms of the number of individuals in the catch, it was in November, with an average of 463 individuals. Dasyatis pastinaca was the most numerous species in the catch; with an average of 1949

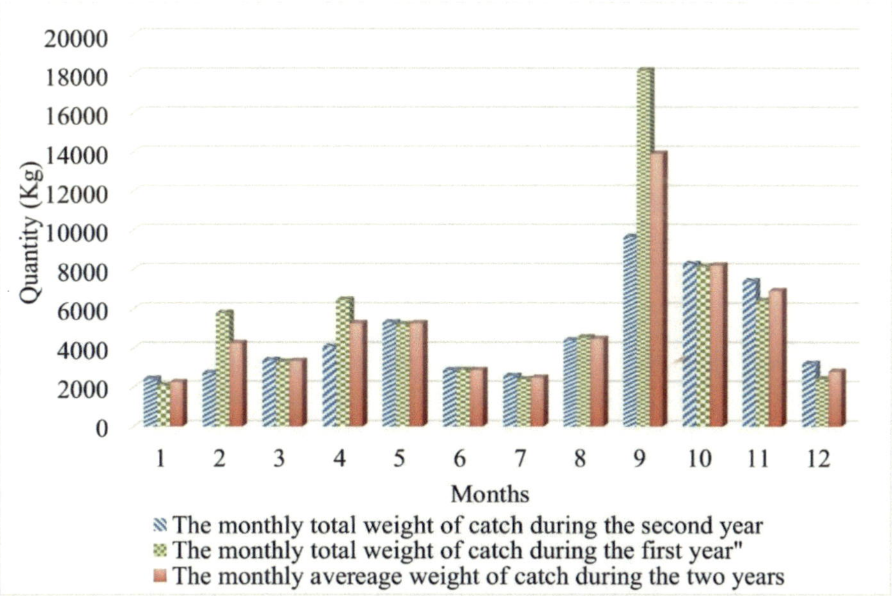

Figure 3.
Distribution of the total weight of cartilaginous fish caught by month from Syrian marine waters during two consecutive years (2015–2016).

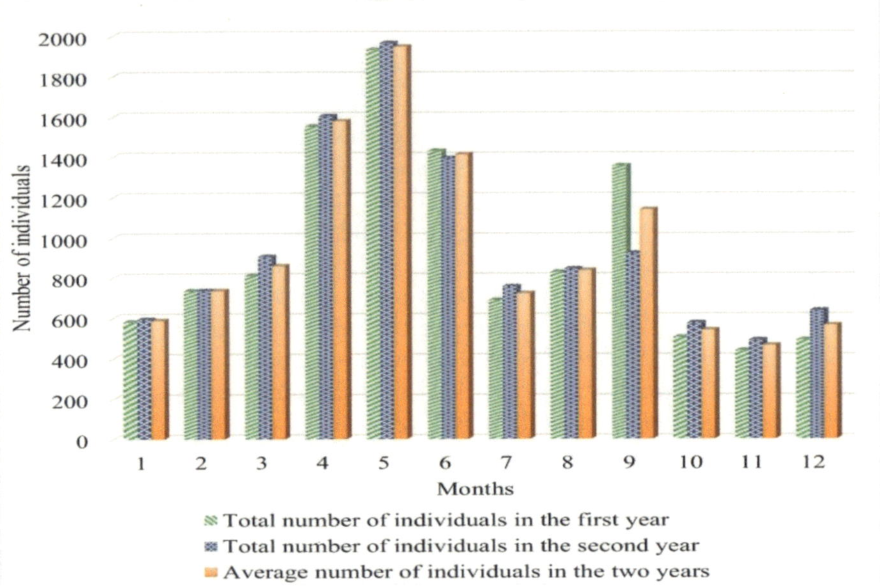

Figure 4.
Distribution of the total monthly number of cartilaginous fish caught from Syrian marine waters during two consecutive years (2015–2016).

individuals, followed by Glaucostegus cemiculus, with an average of 1652 individuals. Seasonal changes in the catch of shark species on the Syrian coast are consistent with bycatch studies of sharks in the eastern Mediterranean [35], which recorded high rates of catch in the months of April (spring) and September (autumn), due to the water temperature preferences, which force the sharks to go to the depths toward the cooler water masses, especially in the regions tropical and subtropical, at these depths sharks are less likely to be caught by longlines, reflecting lower catch rates.

The by-catch of cartilaginous fish on the Syrian coast is rarely returned to the sea, especially in the last 10 years due to the high fish prices and the general decline in the amount of fish caught (bony and cartilaginous), unlike many Mediterranean countries that exclude part of the by-catch of cartilaginous fish. Commercial fishing of target species such as *C. plumbeus* and *Mustelus spp* is seasonal and peaks in spring and summer when the species move to shallower waters. The studies [13, 14] showed regular seasonal movement toward the shore, which may be associated with the seasonal reproductive cycle. There are several species of sharks and rays of different sizes caught, but mainly the juveniles are bycatch in coastal fisheries (**Table 1**). These classes include mainly Triakids, Dasyatids, *Aetomylaeus bovinus*, and the small juveniles individuals of Carcharhinids. For deep fisheries (trawl nets and longlines) many sharks are caught: *Scyliorhinus canicula*, *Galeus melastomus*, *Squalus plainvillei*, *Centrophorus granulosus*, *Mustelus spp*., and some rays are common in the catches. After 2012, due to the high prices of bony fishes and reduction of fuel, all rajids and other chondrichthyans are landed to be marketed.

The data mentioned in **Table 1** indicate the seriousness and great damage to which stocks of cartilaginous fish are exposed in the Syrian marine waters, as a large percentage of the caught individuals are juvenile, sometimes reaching 100 per cent in many species (*Isurus oxyrinchus*, *Alopias superciliosus*, *Oxynotus centrina*, *Mobula mobular*,

Torpedo marmorata, and *T. nobiliana*), and most other species have juvenile individuals above 50 per cent of the total number of individuals caught. The coastal area between Burj Islam and Ras Ibn Hani, North of Lattakia (**Figure 2**) constituted a nursery area for two species of shark, and the coastal area between Latakia and Jableh (**Figure 2**) constituted a nursery area for six species of batoids [13] the above-mentioned. Many of the ecological characteristics in the Syrian marine water are under threat with areas of potentially high cumulative threats overcoming the western and eastern basins and fewer in the southeastern region [36]. The most important causes of current and future threats for the cartilaginous species in the Syrian coast and Levantine basin are habitat loss and degradation, pollution, overexploitation, eutrophication, maritime traffic, invasive alien species, human disturbance, climate change, and bycatch. Juvenile bycatch of commercial species may harmfully affect the future stock and catch levels. Nevertheless, the bycatch of endangered species such as elasmobranchs can have troublesome ecological consequences. These groups of species are susceptible due to their specific biological characteristics [37], especially elasmobranchs that are characterized by their slow growth rate, late maturity, and low fecundity compared to bony fish. Trawling is considered responsible for a large number of elasmobranch bycatches and discards throughout the world. Trawling is prohibited in the territorial marine waters of Syria. This method is problematic and leads to juvenile catches, important discards, and negative impact on the environment. Trammel nets and gillnets are the most frequently used by small Mediterranean fisheries and there is little use of gillnets targeting sharks.

4. Conclusion

Forty-nine cartilaginous fish species include 47 Elasmobranches and two chimaera species has been recorded in the Syrian marine waters. One species (*Pristis pectinata*) that has disappeared from Syrian waters during the last three decades, it is likely that this disappearance was caused by overfishing and the destruction of the environmental habitat due to the increase of various human activities. While previously mainly bycatch, elasmobranch is now being targeted directly by commercial fishing, due to declining catches of bony fishes on the one hand and increasing consumer acceptance of cartilaginous fishes on the other. The fish catch of cartilaginous fish (Elasmobranches) represents an average of 3.68% of the total catch of Syrian marine fish during 5 years of the field study (2011–2016). Trawlers, pelagic longlines and seines appear to be the greatest threat to elasmobranch species. Sharks of the families Hexanchidae, Carcharhinidae and Scyliorhinidae and batoids of families Dasyatidae, Rajidae, Glaucostegidae, and Rhinobatidae constitute the main and commercial part of the cartilaginous fish catch on the Syrian coast. There is an important habitat for much cartilaginous fish in the Syrian marine waters, and there are sustainable communities for six species of cartilaginous fish (two sharks and four batoids). Eleven species of Elasmobranches (six sharks and five batoids) are exposed to overfishing in the Syrian marine waters. There has been a significant decline in catches in terms of both quantity and body size (total length and disc width) of individuals caught over the past 10 years [13, 14]. The coastal area between Burj Islam and Ras Ibn Hani, North of Lattakia, constituted a nursery area for two species of shark, and the coastal area between Lattakia and Jableh constituted a nursery area for six species of batoids.

5. Recommendations

Capture of 11 species of cartilaginous fish should be banned, six species of sharks: *H. Perlo*, *Centrophorus uyato*, *I. oxyrinchus*, *M. mustelus*, *C. Plumbeus*, and *H. griseus*, especially during the period from the beginning of winter to the end of summer, and five species of batoids: *R. clavata*, *Dipturus oxyrinchus*, and *D. tortonesei*, *P. violacea* and *R. marginata*, especially during the spring and summer seasons, by-caught species should be released. Preventing catching of the three migratory species; two sharks: *A. superciliosus* and *Isurus oxyrinchus* and one of batoids: *Mubula mobular* taking steps to collect reliable statistics on elasmobranch landings and bycatch should be a priority for shark conservation, additionally, conducting more biological studies on species that have not been studied in Syrian marine waters.

Conflict of interest

The authors declare no conflict of interest.

Author details

Adib Saad[1*] and Hasan Alkusairy[2]

1 Al Andalus University, Tartus, Syria

2 General Commission for Fisheries Resources and Aquatic Organisms, Jablah-Lattakia, Syria

*Address all correspondence to: adibsaad52@gmail.com; adib.saad53@au.edu.sy

References

[1] Last PR, Seret B, Naylor GJP. A new species of guitarfish, Rhinobatos borneensis sp. nov. with a redefinition of the family-level classification in the order Rhinopristiformes (Chondrichthyes: Batoidea). Zootaxa. 2016;**4117**(4):451-475

[2] Nelson J, Grande TC, Wilson MVH. Fishes of the World. 5th ed. United States of America: Wiley; 2016. p. 707

[3] Roskov Y, Ower G, Orrell T, Nicolson D, Bailly N, Kirk PM, et al., editors. Species 2000 & ITIS Catalogue of Life, 25th March 2019. Leiden, the Netherlands: Naturalis; 2020. pp. 2405-8858. Available from: www.catalogueoflife.org/col

[4] Ebert DA, Stehmann MFW. Sharks, batoids, and chimaeras of the North Atlantic. In: FAO Species Catalogue for Fishery Purposes. No. 7. Rome: FAO; 2013

[5] Skliris N. Past, present and future patterns of the thermohaline circulation and characteristic water masses of the Mediterranean Sea. In: Goffredo S, Dubinsky Z, editors. The Mediterranean Sea. Its History and Present Challenges. Heidelberg: Springer Science; 2014. pp. 29-48

[6] Golani D. An updated checklist of the Mediterranean fishes of Israel, with illustrations of recently recorded species and delineation of Lessepsian migrants. Zootaxa. 2021;**4956**(1):1-108

[7] La Grande-Syrie AJ. Diachronie d'une construction géographique (1912-1923). Revue des mondes musulmans et de la Méditerranée. 2017;**141**:33-35

[8] Saad A. Check-list of bony fish collected from the coast of Syria. Turkish Journal of Fisheries and Aquatic Sciences. 2005;**5**:99-106

[9] Ali M. An updated checklist of marine fishes from Syria with an emphasis on alien species. Mediterranean Marine Science. 2018;**19**(2):388-393

[10] Eschmeyer W N, Fricke R, van der Laan R. Catalog of Fishes classification. 2018. Available from: https://www.calacademy.org/scien-tists/catalog-of-fishesclassification [Accessed: September 20, 2022]

[11] Froese R, Pauly D, editors. FishBase. World Wide Web electronic publication; 2019. Available from: www.fishbase.org Accessed: August 2019

[12] Fricke R, editors. Eschmeyer's Catalog of Fishes: References. California Academy of Science. 2020a. Available from: http://researcharchive.calacademy.org/research/ichthyology/catalog/fishcatmain.asp. [Accessed: September 8, 2022].

[13] Alkusairy H. The Study of Catches Composition (Quantitative and Qualitative) of Chondrichthyes, and Reproductive Biology and Diet of Dipturus oxyrinchus in Syrian Marine Waters [PHD Thesis]. Syria, Lattakia: Tishreen University, Faculty of Agriculture; 2019. (Published in Arabic, with abstract in English)

[14] Alkusairy H, Saad A. Species composition, diversity and length frequency of by-catch sharks from the Syrian coast. International Journal of Research Studies in Zoology. 2018;**4**(1): 11-21. DOI:10.20431/2454-941X.040100

[15] Ali M, Saad A, Kurbaj H, Jnaidi S. Biology of Reproductive and Feeding

of Three Commercial Shark Species in Syrian Marine Water. Syria: The Higher Commission for Scientific Research. Final report; 2013. p. 57 (in Arabic)

[16] Ali M. Reproductive Biology and Feeding of Two Cartilaginous Fish *Mustelus Mustelus* and *Rhiobatos Cemiculus* in Syrian Marine Waters, [PHD Thesis]. Syria, Lattakia: Tishreen University, Faculty of Agriculture; 2009. (In Arabic with abstract in English)

[17] Alkusairy H, Saad A. Some morphological and biological aspects of longnosed skate, *Dipturus oxyrinchus* (Elasmobranchii: Rajiformes: Rajidae) in Syrian marine waters (eastern Mediterranean). ActaIchthyologica Et Piscatoria. 2017;**47**(4):371-383. DOI:10.3750/AIEP/02283

[18] Saad A, Amin A, Alkusairy H, Sabour W. Some reproductive aspects of cartilaginous fish *Raja clavata* L., 1758 (Elasmobranchii: Rajidae) from Syrian coast. Iraqi. Journal of Aquaculture. 2018;**15**(2):103-115. (In Arabic abstract in English)

[19] Alkusairy H, Ali M, Saad A, Reynaud C, Capapé C. Maturity, reproductive cycle and fecundity of spiny butterfly ray, *Gymnura altavela* (Elasmobranchii: Lamniformes: Gymnuridae) from the coast of Syria (eastern Mediterranean). Acta Ichthyologica Et Piscatoria. 2014;**44**(3):229-240. DOI:10.3750/AIP2014.44.3.07

[20] Gruvel A. Les etats de Syrie, richesses marines et fluviales. In: Exploitation actuelle et avenire. Maritimes et colonials: Societe des editions Geographiques; 1931. p. 453

[21] Anonymous Commercial fish species collected by the Korean mission for scientific cooperation in Syria, local report, Ministry of Agriculture. 1976:**76**

[22] Ali M. Taxonomic Economic Study of the Chondrichthyes Fish in Syrian Marine Water [MSc Thesis]. Syria, Lattakia: Tishreen University, Faculty of Agriculture; 2003. (Published in Arabic, with abstract in English)

[23] Saad A, Seret B, Ali M. Liste commentée des Chon-drichthyens de Syrie (Méditerranée orientale). Rapport Commission international Mer Méditerranée. 2004;**37**:430

[24] Saad A, Ali M, Seret B. Shark exploitation and conservation in Syria. In: The Proceedings of the International Workshop on Mediterranean Cartilaginous Fish with Emphasis on Southern and Eastern Mediterranean; 14th–16th October 2006; Istanbul. Turkey: Ed. Turkish Marine Research Foundation (TUDAV); 2006. pp. 202-208

[25] Ali M, Saad A, Reynaud C, Capape C. Occurrence of basking shark, *Cetorhinus maximus* (Elasmobranchii:Lamniformes: Cetorhinidae), off the Syrian coast (eastern Mediterranean) with first description of egg case. Acta Ichthyologica et Piscatoria. 2012;**42**(4):335-339

[26] Saad A, Alkusairy H. Occurrence of mature female of *Etmopterus Spinax* (Chondrichthyes: Etmopteridae) in the Syrian coast (eastern Mediterranean). Advances in Oceanography & Marine Biology. 2018;**1**(1):1-3. DOI:AOMB. MS.ID.000503

[27] Alkusairy H, Saad A. First record of *Leucoraja circularis* (Chondrichthyes: Rajidae) in the Syrian marine waters (eastern Mediterranean). Marine Biodiversity Records. 2018;**11**:5. DOI:10.1186/s41200-018-0140-7

[28] Alkusairy H, Saad A. First record of shagreen ray *Leucoraja fullonica* (Linnaeus, 1758) in Syrian coastal waters (eastern Mediterranean). Cahier de Biologie marine (CBM). 2019;**60**:303-306

[29] Saad A, Basha NA, Tufahha A, Barakat I, Capapé C. First substantiated record of leopard whipray, *Himantura leopard* (Myliobatoidei: Dasyatidae) from the Syrian coast (eastern Mediterranean Sea). FishTaxa. 2021;**19**:5-8

[30] Ali M, Saad A, Ben Amor M, Capape C. First records of the honeycomb stingray, *Himantura uarnak* (Forskal, 1775), off the Syrian coast (eastern Mediterranean), (Chondrichthyes: Dasyatidae). Zoology in the Middle East. 2010;**49**:104-106

[31] Ali M, Saad A, Christian R, Capape C. First records of round fantail stingray *Taeniura grabata* (Chondrichthyes: Dasyatidae) off the Syrian coast (eastern Mediterranean). Zoology in the Middle East. 2013;**59**(2):176-178

[32] Hassan M. Occurrence of large-eyed rabbitfish *Hydrolagus mirabilis*, Chimaeridae, in Syrian waters (eastern Mediterranean). Marine Biodiversity Records. 2013;**6**:e7

[33] Melendez MJ, Báez JC, Serna-Quintero JM, Camiñas JA, IdL F, Real R, et al. Historical and ecological drivers of the spatial pattern of Chondrichthyes species richness in the Mediterranean Sea. PLoS One. 2017;**12**(4):e0175699. DOI:PONE.0175699 10.1371/journal.pone.0175699

[34] Serena F, Abella AJ, Bargnesi F, Barone M, Colloca F, Ferretti F, et al. Species diversity, taxonomy and distribution of Chondrichthyes in the Mediterranean and Black Sea. The European Zoological Journal. 2020;**87**(1):497-536. DOI:10.1080/24750263.2020.1805518

[35] Megalofonou P, Damalas D, Yannopoulos C. Composition and abundance of pelagic shark by-catch in the eastern Mediterranean Sea. Cybium. 2005;**29**:135-140

[36] Coll M, Piroddi C, Steenbeek J, Kaschner K, Ben Rais Lasram F, Aguzzi J, et al. The biodiversity of the Mediterranean Sea: Estimates, patterns, and threats. PLoS One. 2010;**5**(8):e11842

[37] Musick JA. Management of sharks and their relatives (Elasmobranchii). In: Musick J, Bonfil R, editors. Elasmobranch Fisheries Management Techniques. FAO Fisheries Technical Paper. Vol. No. 474. Rome: FAO; 2005. pp. 1-8

Section 3

Management and Conservation Options

Chapter 5

Sharks' Status in the Mediterranean Sea Urgent Awareness is Needed

Mohamed Nejmeddine Bradai, Samira Enajjar and Bechir Saidi

Abstract

Cartilaginous fishes are a very ancient group; Sharks have lived on Earth for about 500 million years, since long before the dinosaurs. They have been able to survive and overcome five mass extinctions since their appearance. They play in fact a key role in maintaining the balance of marine ecosystems. More than 97% of the shark population has disappeared in the last two centuries. Currently, a high percentage is threatened, many are data-deficient. Many menaces face sharks, but fishing pressure seems to be most critical. This chapter focuses on the current status of cartilaginous fishes and progress on conservation measures and actions taken mainly through regional plans.

Keywords: Sharks' status, Mediterranean Sea, conservation, action plans

1. Introduction

Sharks, skates, and rays, collectively referred to as elasmobranchs (Class *Chondrichthyans*), are a very ancient fish group, they have been able to survive and overcome five mass extinctions since their appearance. They play in fact a key role in maintaining the balance of marine ecosystems.

The current chondrichthyans fish fauna of the Mediterranean is relatively diverse, with at least 48 species of sharks, 40 of batoids, and 2 of chimaeras, even if some of them have to be confirmed.

Chondrichthyans grow slowly, mature late, and give generally few babies. Because of these features, they are vulnerable to overexploitation and any anthropogenic disturbance.

More than 97% of the shark population has disappeared in the last two centuries. Currently, a high percentage is threatened, many are data-deficient. Many menaces face sharks, but fishing pressure seems to be most critical.

Several species are currently threatened in the Mediterranean; 53% of 73 species assessed are critically endangered, endangered, or vulnerable according to the IUCN Red List. Thirteen percent are data-deficient to be assessed [1].

Following this bad situation of this fish group, many conventions, RFMOs, and NGOs are dealing with the conservation of cartilaginous fish, and national and regional action plans were developed for this issue. This chapter focuses on the current status of cartilaginous fishes in the Mediterranean Sea and progress on conservation measures and actions taken mainly through regional plans.

2. Tools applying for the conservation and management of sharks in the Mediterranean sea

To counteract the high vulnerability of sharks, international legal instruments applying for the conservation and management of sharks were developped to coordinate the management, conservation, exploration, and exploitation of living resources of the riparian countries, with respect to the protection and preservation of the marine environment.

2.1 The Mediterranean action plan (PAM)

On this regard and concerning environemental issue, the Mediterranean Action Plan (PAM), in the context of the United Nations Environment Programme (UNEP), established the "Convention for the protection of the marine environment and the coastal region of the Mediterranean Sea" (Barcelona convention). Among their protocols, the Protocol concerning specially protected areas and biological diversity (SPA/BD) is of utmost importance for the protection of sharks.

2.2 The general fisheries commission for the Mediterranean (GFCM)

In the field of fisheries, management and conservation are implemented within the framework of the General Fisheries Commission for the Mediterranean (GFCM). It is a Regional Fisheries Management Organization whose main objective is the conservation of living marine resources and its sustainable use, as well as the development of aquaculture in the Mediterranean and the Black Sea in a soustainable way to preserve the environment. The Commission adopted binding recommendations for the conservation and the management of fisheries in the region within its scope. In particular, its measures may relate, for example, to the regulation of fishing methods, fishing gear and minimum landing size, the establishment of open and closed fishing seasons and areas, and the control of fishing effort.

The GFCM plays a decisive role in coordinating the efforts of governments to effectively manage fisheries at the regional level, in accordance with the FAO Code of Conduct for Responsible Fisheries. The Data Collection Reference Framework (DCRF) for the collection and submission of data relating to fisheries in the GFCM area was endorsed by the SAC (Scientific Advisory Committee) as an instrument to assist Contracting Parties to comply with existing recommendations for the collection and submission of fisheries data to GFCM.

2.3 The International Commission for the conservation of the Atlantic Tuna (ICCAT)

Beside the GFCM, the International Commission for the Conservation of the Atlantic Tuna (ICCAT) is dealing with species fished in association with tuna, such as sharks.

2.4 Other conventions

In addition to the GFCM and ICCAT, the convention on the law of the sea (UNCLOS), the convention on the conservation of migratory species of wild animals

(CMS convention), and the convention on international trade of endangered species (CITES) contain some provisions that are relevant for the sharks issue.

2.5 Action plans for the conservation and management of sharks

Although, non-binding instruments, Action Plans for the conservation and management of sharks, contribute usefully for the protection of these species; at global level, the FAO International Plan of Action (IPOA-Shark) and at regional level, the Action Plan for the conservation of the cartilaginous fishes (Chondrichthyans) in the Mediterranean, managed by a Secretariat of UNEP Regional Seas.

Recently and with the aim of protecting the three Mediterranean angel sharks, the second most threatened family of sharks in the world and listed in annex II, the *Shark Trust* and the IUCN Shark Specialist Group (SSG), with regional and international experts developed in 2019 for the Mediterranean Angel Sharks, the Regional Action Plan (RAP).

Taking into accompt recommandations of the International Action Plan (FAO IPOA-Sharks) and the UN Fish Stocks Agreement, the Barcelona Convention, in the frame of the Mediterranean Action Plan, elaborated the Action Plan for the conservation of the cartilaginous fishes (Chondrichthyans) in the Mediterranean [2].

The Action Plan constitutes a mi-term regional strategy that should be updated each 5 years based on an evaluation of their implementation at regional and national levels. The last update dates back to the year 2020 [3].

The objectives of this update are (1) Assess the implementation of the Action Plan in the Mediterranean sea at regional and national levels and (2) Propose a draft updating for the Action Plan.

2.6 The International union for the conservation of the nature (IUCN)

The International Union for the Conservation of the Nature (IUCN) is recognized at international level to provide scientific advice on the conservation status of living species. Through its "Red List of Threatened Species," updated as far as it is possible, the species are classified in nine categories, from Extinct (EX) till Not evaluated (NE). Species that can be considered as threatened are those classified as Critically Endangered (CR), Endangered (EN), and Vulnerable (VU). Criteria used for this red list are used also when assessing status of species and examining proposals for amending Annexes II and III of the SPA/BD Protocol.

3. Progress on shark conservation: review of the implementation of the shark action plans

This review is based mainly on the report of the implementation of the Action Plan for the conservation of cartilaginous fishes (Chondrichthyans) in the Mediterranean (2014–2019) in the frame of its update [4]. It deals with international implementation of the IPOA shark and other international tools, SPA/RAC implementation, national implementation (SPA/RAC focal points and experts' thoughts) and on bibliographic research.

To update the AP, a large consultation with national focal points of SPA/RAC and organims involved in the conservation of sharks at regional level was undertaken via online questionnaire of the Barcelona convention (IG.23/1, CoP20, 2017). The questionnaire was sent by the SPA/RAC to their Focal Points, Convention's Secretariats, the Action Plan Partners, and some experts.

A bibliographic review was also done to collect information on reaserch progress on the issue and new regulations at regional and national level applicable to sharks conservation. Actions planned for the period 2014–2019 were described and analyzed to understand and to review mainly not achieved and ongoing actions.

The implementation focuses on that of the Action Plan for the conservation of cartilaginous fishes (Chondrichthyans) in the Mediterranean at national and international level, by SPA/RAC and comments of RFMOs working on shark conservation issues.

Sixteen responses were received; 10 from SPA/RAC focal points and one response from the IUCN. Five responses were received from experts from four countries.

3.1 Sharks landings through Mediterranean sea

Sharks are generally caught incidentally, while are targeted in some marine areas by small scale vessel. They represent about 1.7% of the total Mediterranean halieutic landings (average for the years 2016–2020) [5]. A decline in cartilaginous species landings has been observed while fishing effort has generally increased. According FAO statistics of elasmobranchs, the catches show a decreasing trend from about 20,000 tons in 2009 to about 12,000 tons in 2020 (**Figure 1**).

During these two decades, the major elasmobranch-fishing countries within the Mediterranean Sea are Libya (since 2009) and Tunisia (**Figure 2**). Italy and Turkey, known to be the major elasmobranch-fishing countries within the Mediterranean in the 1980s, register a dramatic decrease in catch (**Figure 2**).

Statistical information for many priority species of the GFCM area or of the annexes II and III of SPA/RAC is absent as *Mobula mobular, Carcharodon carcharias,*

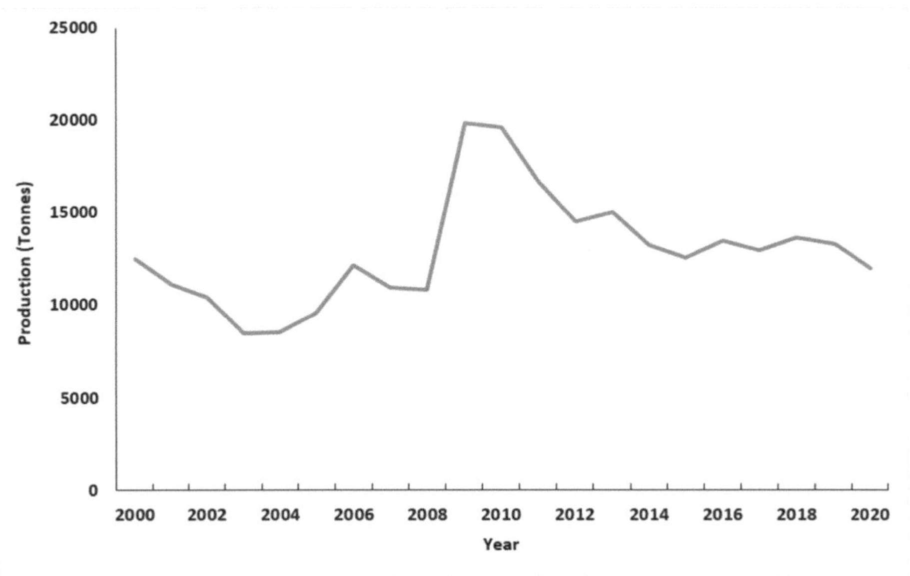

Figure 1.
Mediterranean trends of elasmobranchs catches from 2000 to 2020 according to FAO statistics.

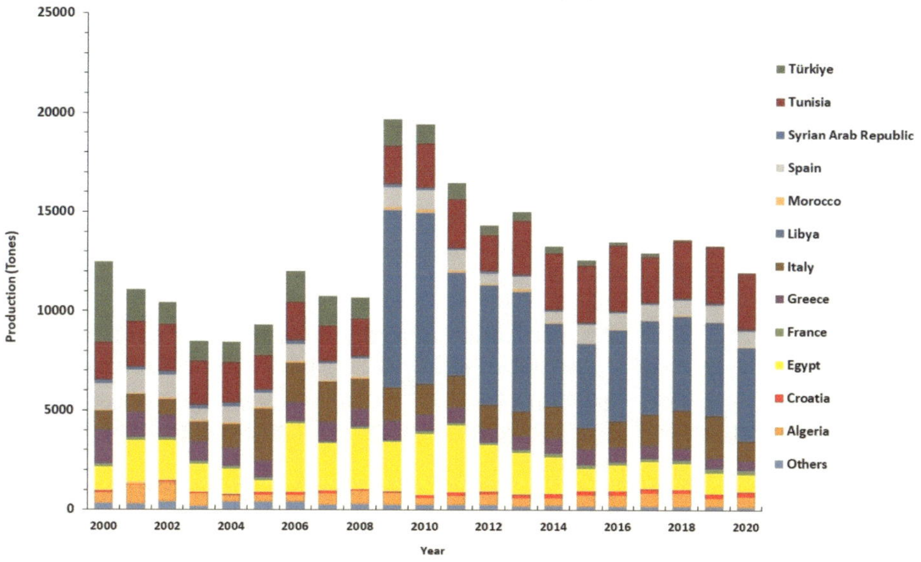

Figure 2.
Contribution of some countries in the Mediterranean elasmobranch production according to FAO statistics from 2000 to 2020.

Gymnura altavella, Raja miraletus... The species of rays and sharks are usually grouped together in the same statistical category and not identified by species.

3.2 Conservation status of cartilaginous fish (Chondrichtyens) in the Mediterranean Sea

Historically, the Mediterranean Sea is a biodiversity hot spot for cartilaginous species. Currently, the region has the highest percentage of endangered sharks in the world. This fish group is in fact vulnerable to overexploitation and any anthropogenic disturbance because of their biological characteristics (very slow growth, late sexual maturity, low fecundity). Several species are now threatened. The main menaces are:

- Fishing pressure and use of nonselective fishing gears;
- Fin trade "fining";
- pollution and habitats degradation;
- Other uses.

Experts estimated that about 97% of Mediterranean elasmobranchs population, expressed in number and catch weight, was lost during the last 200 years [6].

A significant decline in species richness has been confirmed recently due to increasing of mentioned menaces [1].

According to the last elasmobranchs assessment made by the IUCN [1], there is no sign of improvement in the status of Mediterranean Chondrichthyans; on the contrary, the situation seems to be worsened. The rate of threatened species increases from 42.25 to 53% during decade (2007–2016).

A decrease by 8% of data-deficient species between IUCN assessment of 2007 and that of 2016 shows the increasingly interest of scientists on this issue [1–7].

3.3 Trainings/multilingual regional and national field identification guides and sheets

In the context of developing training for capacity building at national and regional level, mainly in following topics: taxonomy, biology, and ecology, some trainings and field identification guides were realized:

- The ACCOBAMS-GFCM project on mitigating interactions between endangered marine species and fishing activities (2016–2017) produced a guideline distributed mainly to fishermen and observers titled "Good practice guide for the handling of sharks and skates caught incidentally during pelagic longline fishing."

- SELPAL and RéPAST projects produced a Responsible Fisherman's Guide: Best Practices for Reducing the Mortality of Sensitive Species Incidentally Caught by French Pelagic Longliners in the Mediterranean (in French);

- In the frame of bycatch project "Understanding Mediterranean multi-taxa 'bycatch' of vulnerable species and testing mitigation- a collaborative approach" (2017–2022), jointly implemented by BirdLife (Project coordinator), GFCM, ACCOBAMS, SPA/RAC, IUCN Med, MEDASSET, and WWF and financed by MAVA foundation, many useful documents were elaborated dealing with collect of data on bycatch, species identification (including sharks), bycatch of sharks:

    - o Monitoring the incidental catch of vulnerable species in Mediterranean and Black Sea fisheries: Methodology for data collection (http://www.rac-spa.org/sites/default/files/doc_bycatch/ca4991en.pdf)

    - o Identification guide of vulnerable species incidentally caught in Mediterranean Fisheries http://www.rac-spa.org/sites/default/files/doc_bycatch/final_identification_guide_20_12.pdf

    - o Pocket identification guides of main vulnerable species incidentally caught in fisheries (in each language of countries involved in the project) http://www.rac-spa.org/sites/default/files/doc_bycatch/medbycatch_pocket_guide_tn_en.pdf

    - o incidental catch of vulnerable species in Mediterranean and black sea fisheries- a review https://www.fao.org/3/cb5405en/cb5405en.pdf

In the frame of this project, virtual training courses on the identification and handling of vulnerable species incidentally caught during fishing operations were organized.

- The FAO produced in 2018 Two documents' *Species Photographic Plates* for sharks and another for skates, rays and chimaeras [8, 9].

- A manual on the identification of elasmobranchs species was produced by the NGO ASCOB-Syrtis, the SPA/RAC and the INSTM for the need of training courses on taxonomy. http://www.rac-spa.org/sites/default/files/doc_sharks/requins_fr_web.pdf

- In the frame of MAVA Species Knowledge Project (2019–2021), coordinated by SPA/RAC in collaboration with ACCOBAMS, BirdLife, MEDDASSET and IUCN, ASCOB-Syrtis executed an action titled: "Status of elasmobranchs, sea turtles and cetaceans in purse seine and surface longline fisheries in the gulf of Hammamet (Tunisia) and produced a manual for mainly identification of sharks and batoids." https://aquadocs.org/handle/1834/41741

3.4 Submission of catch, bycatch, and discard data to the GFCM

The incidental capture of vulnerable species in fisheries represents a key conservation issue for a number of taxonomic groups (i.e., sea turtles, marine mammals, seabirds, and elasmobranchs). Reliable information and reporting, as well as efficient measures, will allow scientists and decision-makers to obtain a more complete overview of the situation and set priorities for management action.

Several goals were set by the AP on sharks toward reaching more sustainable fisheries in the Mediterranean, notably by decreasing the interaction with shark's species mainly bycatch. For this, contracting parties to the Barcelona convention were requested to develop program to gather information on the catch rate, the amount of bycatch, marine litter and discard, fishing gears used, and some biologic parameters.

The GFCM recommends mainly to contracting parties, throw GFCM/36/2012/3 recommendation, amended by GFCM/42/2018/2 recommendation:

- To inform on fishing activities (landings, bycatch, release and discard);

- To ban catch of the 24 species of shark listed in annex 11 of the SPA/BD protocol.

Responses were received from nine countries concerning implementation of this action: two responded by yes, three by no, and the action is under development for three while it is not applicable for one country. Even if few responses were received, the implementation seems to be underway with poor progress.

A new analysis of information collected from 2000 to 2020 on incidental catch of sharks [5] DOI: 10.4060/cb2429en reported incidental catch of the main elasmobranch species in the Mediterranean Sea (**Figure 3**).

Very few species of annexes II and III were reported in this analysis indicating that submission of shark bycatch data to the GFCM is not well implemented.

Longliners and bottom trawlers are by far the vessel groups with the greatest impact on conservation priority elasmobranch species in the whole region. Small-scale vessels and pelagic trawlers generate a minor impact on these vulnerable species. Purse seine seems to have the lowest impact (**Figure 4**) [5].

3.5 Legal processes

3.5.1 Species protection

Management of elasmobranchs has been widely emphasized over the past 23 years in a precautionary way. Since 1999, the international action plan of the FAO has undertaken various conservative actions to improve exploitation management of this fish group. In the period of 2010–2011, the GFCM adopted ad hoc mitigation measures concerning bycatch of some pelagic sharks such as thresher sharks, shortfin mako

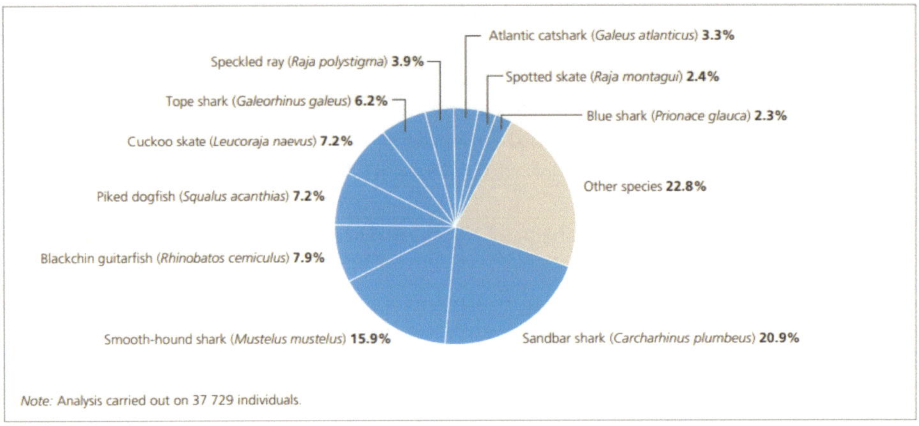

Figure 3.
Reported incidental catch of the main elasmobranch species in the Mediterranean Sea, 2000–2020.

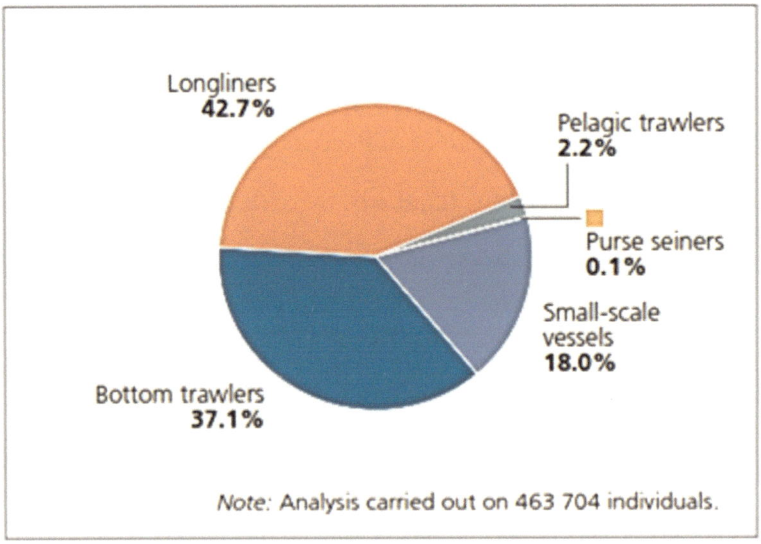

Figure 4.
Bycatch of elasmobranchs reported by vessel group in the GFCM area (2000–2020) [5].

sharks, and hammerhead sharks. Then in 2012, the GFCM banned finning practices and capture of threatned sharks and rays of the Appendix II of the prtocol on Special Protected Areas and Biodiversity of the Barcelona Convention.

Species protection is a major objective of the Action Plan. The urgent provision of legal protection status for endangered species is listed as the first priority; CPs were asked to establish strict legal protection for species listed in Annex II and GFCM Recommendation through national laws and regulations as soon as possible.

Although 80% of CPs who responded to the questionnaire reported that they had protected species or had measures under development, implementation has been incomplete, with 10% of CPs reporting no action. All the Annex II species have been included in legislation of three countries. Some countries protect a smaller number of Annex II species.

Overall, the level of protection granted to the chondrichthyan fish species listed in Annex II of the Protocol continues to be disappointingly incomplete.

3.5.2 Finning prohibition

The International Commission for the Conservation of Atlantic Tuna (ICCAT) and the General Fisheries Commission for the Mediterranean (GFCM) recommend respectively in 2004 and 2005 full utilization of these animals and that landed fins do not exceed 5% of the sharks caught (in weight).

Later, in 2012, The GFCM banned finning practices following recommendation GFCM/36/2012/3 concerning conservation of elasmobranchs in the GFCM area.

This former recommendation was later, in 2018, amended by Recommendation GFCM/42/2018/2 mandating that sharks must be landed with their fins attached to the body. This measure reinforces the ban of "finning" consisting on cutting of fins and discarding the body.

Following responses to the questionnaire, only 50% of the parties to the GFCM Support finning prohibition by enacting national regulations and monitoring their implementation and enforcement and 10% (one CP) have a project on this matter under development.

Three parties run into difficulties to implement this action due regulation, administrative and technical matters.

3.5.3 Protection of critical habitats for chondrichthyans

The inquiry shows a poor protection of critical habitats for elasmobranchs in the Mediterranean. An indirect protection appears may be through the recommendation GFCM/42/2018/2 of the GFCM dealing with the reduction of the fishing effort of trawlers in coastal areas aiming protection of coastal elasmobranchs.

3.6 Studying programs

3.6.1 Scientific publishing effort

The parties to the Barcelona convention, having adopted the action plan on cartilaginous fish, were asked to undertake continuously monitoring programs to elucidate mainly the status of Mediterranean endemics and Data-Deficient or Near-Threatened species. To have an idea on the implementation of this recommendation, we undertake a review and an analysis of the bibliography on this matter during the last decade.

Three hundred papers on elasmobranchs appeared in 2012–July 2022 period in Mediterranean Sea with an average of about 30 papers/year. They cover different areas and many topics. The most papers came from the Eastern Mediterranean followed by the central Mediterranean Sea (**Figure 5**).

The analysis shows also that the main topics of papers concern biology, ecology, and fisheries (**Figure 6**). Few papers were published on stock assessment. Studies on shark systematic and genetics increased a lot compared with last decades before 2012. This interest is mainly due to the known identification problems in elasmobranchs. The increase of papers on fisheries and mainly bycatch seems to be related to the implementation of programs on this topic to reduce this phenomenon. Studies concerning this topic, ranked second, represent about 25% of the available papers.

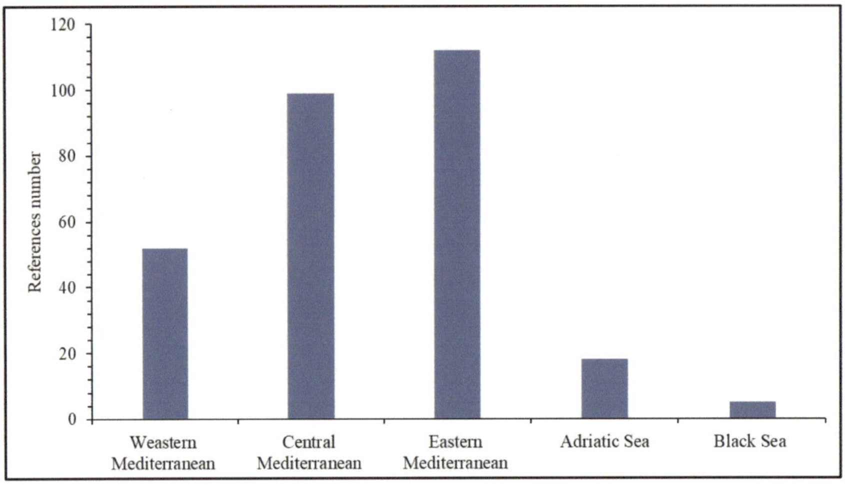

Figure 5.
Geographic distribution of elasmobranchs paper in the Mediterranean Sea between 2012 and 2020.

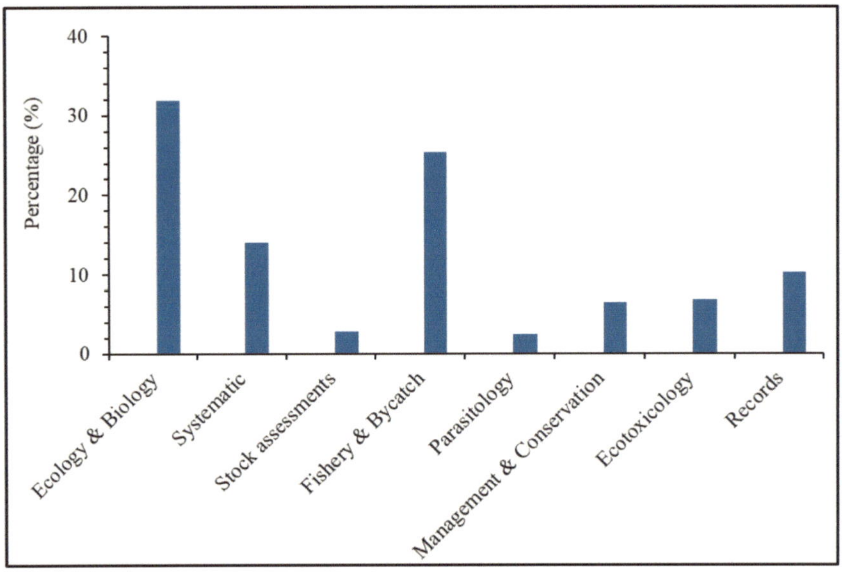

Figure 6.
Distribution of elasmobranchs paper by topic in the Mediterranean Sea between 2012 and 2020.

3.6.2 The Mediterranean large elasmobranchs monitoring

The MEDLEM program was launched in 1985 [10] and adopted by FAO-GFCM and UNEP-RAC/SPA respectively in 2005 and 2009. The "Shark Specialist Group" of the International Union of the Conservation of the Nature and the "European Elasmobranchs Association" are partners [11]. The main goal of the program is to evaluate the elasmobranchs bycatch recording the incidental catches, sightings, stranding, and historical bibliographic references.

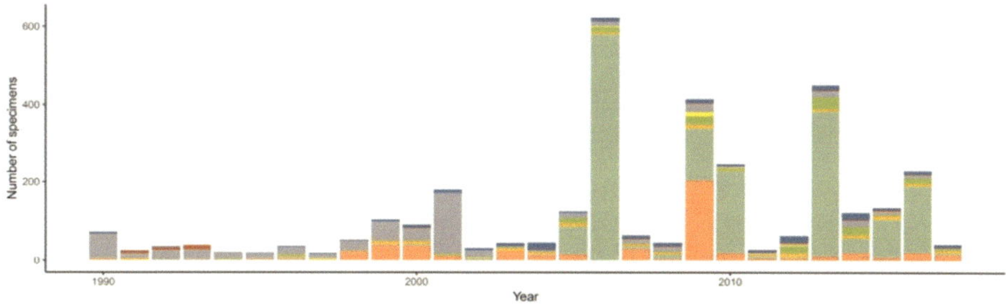

Figure 7.
Number of records of the main species registered between 1990 and 2017.

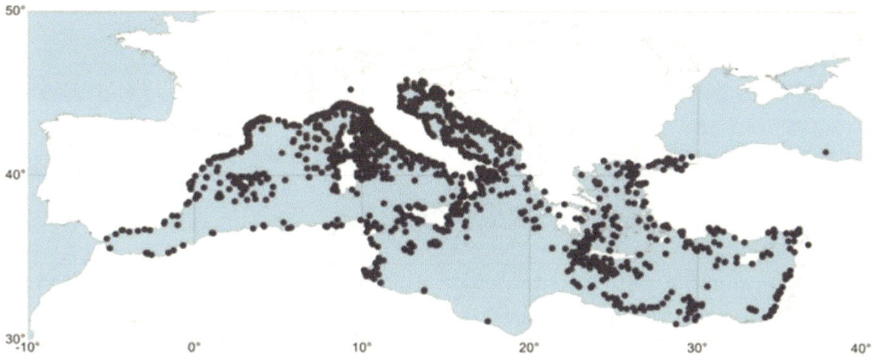

Figure 8.
Spatial distribution of all MEDLEM records [12].

Twenty different countries participate in the MedLem program and promote input and shared access to the database under the appropriate protocol. Records are clearly increasing (**Figure** 7).

The MEDLEM database could be a useful tool for organizations involved at national and international level to manage conservation of elasmobranchs and Mediterranean biodiversity.

Observation effort is more important in in the northern sectors than in the south-eastern ones. No records of some species in one of these regions does not mean their absence (**Figure 8**).

Actually, the database is hosted by the GFCM server and continues to be enriched. To contribute more on the conservation efforts, the database should be open to all elasmobranch's species and not only large ones.

3.6.3 Critical habitats for chondrichthyans

Critical habitats should be identified for conservation purposes. In fact, a big lack of knowledge on critical habitats for this group was noted in the Mediterranean [13]. However, some studies confirm the presence of nursery and spawning areas for some species such as in the Gulf of Gabes (GSA 14) in Tunisia. These critical habitats concern mainly the sandbar shark, smooth hounds, and guitarfishes [14–17]. The Tunisian waters provide also a nursery area for the white shark Carcharodon carcharias (center of Tunisia) [13]. Since the Gulf of Gabès is an important fishery area in Tunisia and in the aim to have valuable knowledge on this kind of area, many biologic parameters should be determined. It should also be noted that aggregations of basking shark have been reported in the Balearic area, the Adriatic, and the Tyrrhenian Sea [18].

Considering the utility of this kind of habitats for conservation purposes, the AP on cartilaginous required parties to proceed with inventorying and mapping such habitats and to ensure their legal protection.

Among 10 responses received on this matter, two parties (20%) inventoried critical habitats, but they are not yet under protection such as nursery areas in the Gulf of Gabes. More work is needed to delimit such nurseries. Three CPs (30%) did not do so, for four countries (40%), investigations are under development, and for one CP (10%), the action is not applicable.

3.6.4 Data collection programs

From 2010 to 2013, the GFCM undertook a research program to evaluate the status of elasmobranchs and to propose measures reducing populations decline. Then, other programs were launched. The following programs should be mentioned:

3.6.4.1 ACCOBAMS-GFCM project (2015–2016)

This project dealing with bycatch evaluation and mitigation measures, aimed to enhance the conservation of endangered marine species, such as cetaceans, sea turtles, elasmobranches, and seabirds, and to promote responsible fishing practices in the Mediterranean through six pilot actions in France, Spain, Morocco, and Tunisia.

For Tunisia, the project was developed in Zarzis (in the Gulf of Gabes) and focusing on longline fishery, including some experiments with circle hooks and change of baits. Below, the link for the final report: https://www.accobams.org/wp-content/uploads/2019/04/07-Tunisie-INSTM-Rapp-final_palangres-Zarzis.pdf

3.6.4.2 MedBycatch project

A bycatch project "Understanding Mediterranean multi-taxa 'bycatch' of vulnerable species and testing mitigation – a collaborative approach" (September 2017–October 2022), implemented by Birdlife International (as coordinator), SPA/RAC, GFCM, ACCOBAMS, MEDASSET, and IUCN-Med and financially supported by the MAVA Foundation in five countries (Morocco, Tunisia, Turkey, Italy, and Croatia) is interested, among other taxa, in cartilaginous fishes.

The project focused on five taxa among them elasmobranchs and five fishing gears: trammel nets, gillnets, bottom trawlers, bottom longline, and purse seine, in more than 45 ports. More than 50 observers were involved, and more than 1500 observations were done.

In Tunisia (GSAs 12, 13 and 14), observation effort deployed in phase 2 (2020–2022) was recorded in **Table 1**. Data analysis shows that elasmobranchs are the taxa the most incidentally caught (**Figure 9**) [18].

3.6.4.3 MEDITS project (Mediterranean International Trawl survey)

The MEDITS survey program (International bottom trawl survey in the Mediterranean) intends to produce basic information on benthic and demersal species in

DOI: http://dx.doi.org/10.5772/intechopen.108162

Fishing gears	*Onboard observations*	*Questionnaires*	*Self-sampling*
OTB	214	257	
SSV	632	1005	27
Purseseiners	38	127	1

Table 1.
Observation effort.

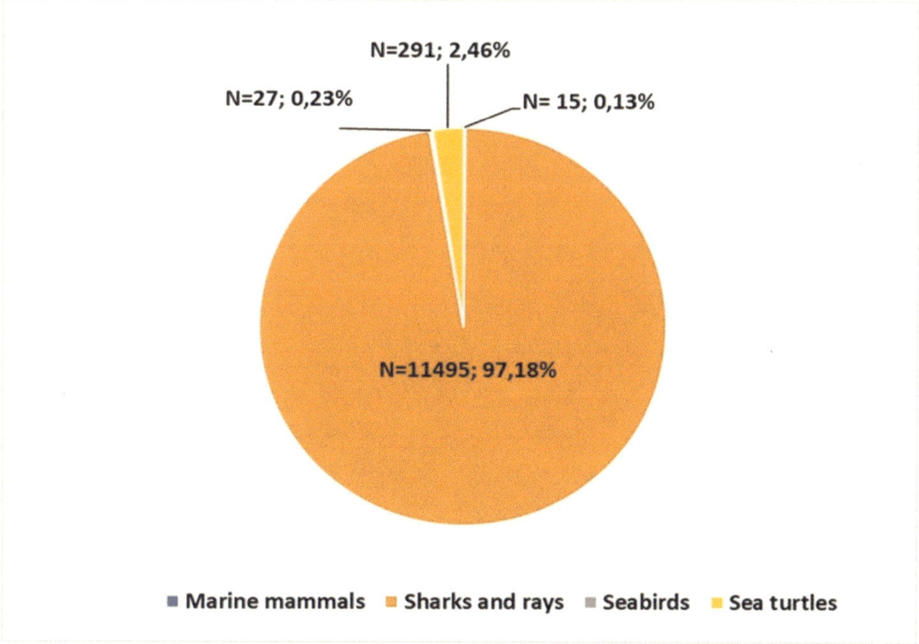

Figure 9.
Total number (%) of bycaught individuals in Tunisia.

terms of population distribution as well as demographic structure, on the continental shelves, and along the upper slopes at a global scale in the Mediterranean Sea, through systematic bottom trawl surveys.

The program aims at conducting coordinated surveys from bottom trawling in the Mediterranean Sea. The surveys intend to include as much as possible all the trawlable areas over the shelves and the upper slopes from 10 to 800 m depth off the coasts of the partner countries. Since 2002, the MEDITS survey is included in the European regulation related to the collection of fishery data.

This survey continues to collect data on many taxa among them cartilaginous fishes. This program needs to be extended to the North African and Levantine countries.

3.6.4.4 MAVA species knowledge project (2019–2021)

The Project aims to improve knowledge on species and habitats affected by unsustainable fishing practices through nine small Projects covering all megafauna species executed by NGOs (elasmobranchs, sea turtles, birds, and cetaceans). Among them two in the Strait of Sicily-Tunisia subregion. **Figure 10** gives a picture of the density map of elasmobranchs in Sicily Canal and Tunisian plateau.

3.6.5 Data submission to FAO and GFCM

3.6.5.1 Data on pelagic shark catches

Pelagic sharks are protected mainly under the Convention on Migratory Species (CMS) and ICCAT recommendations. The inquiry, carried out for the period 2014–2019, shows that few data on this shark group were submitted to the FAO and the GFCM; one country among eight did so.

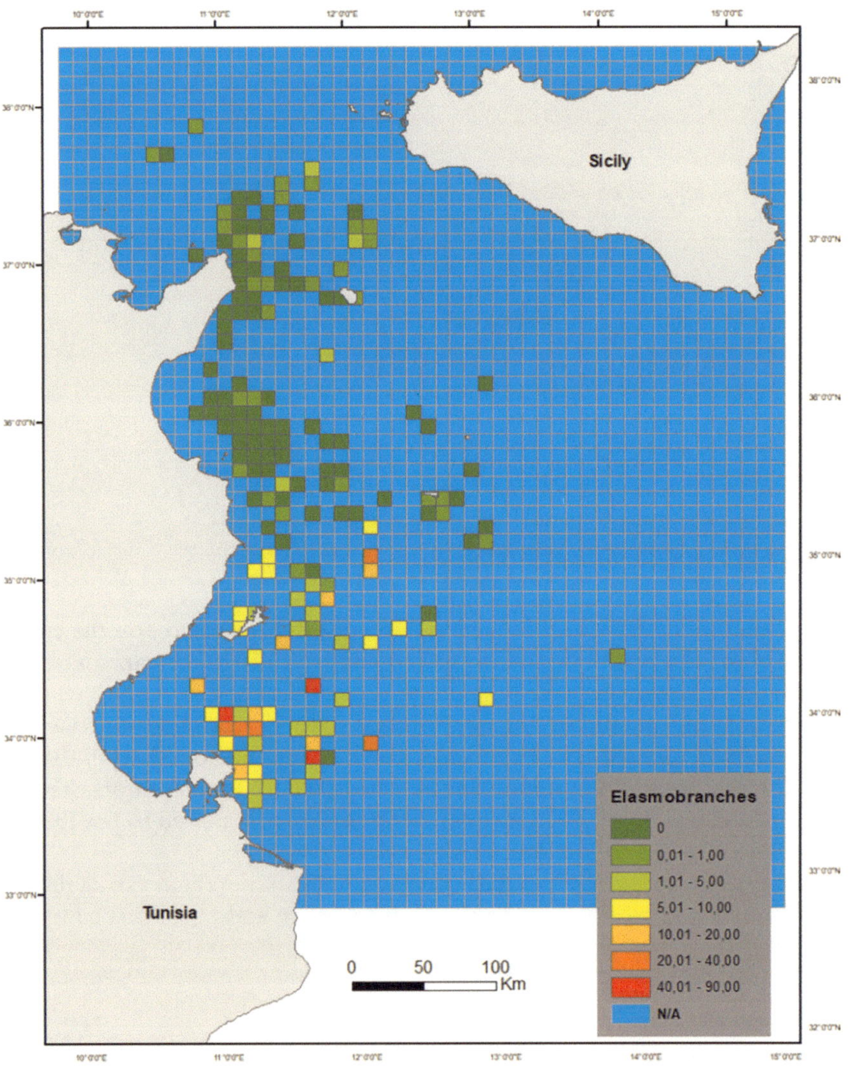

Figure 10.
Catch per fishing day of elasmobranchs per 10 km by 10 km raster (Strait of Sicily and Tunisian plateau).

3.6.5.2 Collection and submission data from coastal fisheries

Few countries (four from nine) contributed to such programs for the period 2014–2019.

3.6.6 Capacity building

GFCM, some FAO project, and other organizations (ACCOBAMS, SPA/RAC …) support expert participation in RFMO and other relevant meetings, training courses, and workshops, to share expertise and build capacity for data collection, stock assessment, and bycatch mitigation. This action seems to be well implemented and should be more improved by supporting experts and students to participate mainly to specific training courses on species identification, data collection, and data analysis.

3.7 Management and assessment procedures

3.7.1 National shark plans

The Mediterranean Action plan recommends to contracting parties the elaboration of national action Plans. The role of SPA/RAC is fundamental for constantly updating the MAP and for stimulating Mediterranean countries to produce their own National Plans.

Until 2021, only countries elaborated National Action Plan for the conservation of sharks. Three Mediterranean.

4. Proposals of priorities to enhance implementation of action plans on shark conservation

This review shows that regional action plans and recommendations of regional conventions and RFMOs seem to be poorly implemented. Since the future of marine biodiversity in the Mediterranean and of sustainable fishery depends a lot on elasmobranchs, more successful conservation of this fish group should be ensured, and urgent awareness is needed focusing mainly on the following priorities:

- Increase training and awareness of fishermen and controllers of fisheries on protected species and develop an effective control of fisheries.

- Improve shark conservation in multi-taxa approach, mainly for:

    - Bycatch reducing mainly through species release;

    - Mapping and monitoring of critical habitats.

- Improve data collection at sea and at land for a global map of species distribution using data obtained in all regional projects;

- Improve collect of elasmobranches landing statistics;

- Strongly ban fishing and landing of vulnerable species (listed in annex II). The GFCM binding recommendation on this issue should be applied since legislative process is long at national level;
- Develop national red lists for elasmobranches
- Improve studies on stock assessment, in fact, analysis of action plans implementation on this issue shows many gaps. To develop this research field, experts should focus on (1) some biologic parameters, (2) knowledge on fisheries, (3) valuable and available statistics, (4) taxonomy and shared stocks, (5) studies on migration and exchange between populations, and (6) choosing better evaluation methods.

Author details

Mohamed Nejmeddine Bradai[1*], Samira Enajjar[1] and Bechir Saidi[2]

1 Carthage University, National Institute of Science and Technology of the Sea, R16INSTM02 Marine Biodiversity Laboratory, Sfax, Tunisia

2 Kairouan University, Faculty of Sciences and Techniques of Sidi Bouzid, Sidi Bouzid, Tunisia

*Address all correspondence to: mednejmeddine.bradai@instm.rnrt.tn

References

[1] Dulvy NK, Allen DJ, Ralph GM, Walls RHL. The Conservation Status of Sharks, Rays and Chimaeras in the Mediterranean Sea. Malaga, Spain: IUCN; 2016

[2] UNEP-MAP RAC/SPA. Action Plan for the Conservation of Cartilaginous Fishes (Chondrichthyans) in the Mediterranean Sea. Tunis: Ed RAC/SPA; 2003. p. 56

[3] SPA/RAC–UN Environment/MAP. In: Bradai MN, editor. Action Plan for the Conservation of Cartilaginous Fishes (Chondrichtyans) in the Mediterranean Sea. Tunis: Ed SPA/RAC; 2020. p. 18

[4] UNEP/MAP-SPA/RAC. 5.1Updating of the regional strategy and action plans for the conservation of the Mediterranean Monk Seal, Marine Turtles and Cartilaginous Fishes in the Mediterranean Sea. (UNEP/MED WG.461/7) Conservation of Species and Habitats Project. 2019

[5] FAO. The State of Mediterranean and Black Sea Fisheries 2020. Rome: General Fisheries Commission for the Mediterranean; 2020

[6] Ferretti F, Myers RA, Serena F, Lotze HK. Loss of large predatory sharks from the Mediterranean Sea. Conservation Biology. 2008;**22**: 952-964

[7] Cavanagh RD, Gibson C. Aperçu du statut de conservation des poissons cartilagineux (Chondrichtyens) en mer Méditerranée. UICN, Gland, Suisse et Malaga, Espagne. 2007. p. VI + 39

[8] FAO. Species photographic plates. In: Barone M, Serena F, Dimech M, editors. Mediterranean Sharks. Rome, Italy. 2018

[9] FAO. Species photographic plates. In: Barone M, Serena F, Dimech M, editors. Mediterranean Skates, Rays and Chimaeras. Rome, Italy. 2018

[10] Mancusi C, Baino R, Barone M, Vacchi M, de Sola LG, Morey G, et al. MedLem Database, a Data Collection on Large Cartilaginous Fishes in the Mediterranean Basin 2-6 June 2014. Durban: Sharks International; 2014

[11] Serena F, Bradai MN, Mancusi C, Barone M. The MEDLEM database application: a tool for storing and sharing data about by-catch, sightings and stranding of large cartilaginous fishes in the Mediterranean basin. (Data update to September 2009). Paper presented at the FAO GFCM Transversal workshop on selectivity improvement and by-catch reduction. Sub-Committee on Marine Environment and Ecosystems SCMEE), Sub-Committee on Stock Assessment (SCSA), Sub-Committee on Economic and Social Sciences (SCESS), Tunis, Tunisia, 23-25 September 2009. 2009

[12] Mancusi C, Baino R, Fortuna C, De sola l G, Morey G, Bradai MN, et al. MEDLEM database, a data collection on large Elasmobranchs in the Mediterranean and Black seas. Mediterranean Marine Science. 2020;**21**(2):276-288. DOI: 10.12681/mms.21148

[13] Bradai MN, Saidi B, Enajjar S. Elasmobranchs of the Mediterranean and Black sea: Status, ecology and biology. Bibliographic analysis. Studies and Reviews. General Fisheries Commission for the Mediterranean. No. 91. Rome, FAO. 2012. 2012. p. 103

[14] Bradaï MN, Saidi B, Bouaïn A, Guelorget O, Capapé C. The Gulf of

Gabès (Southern Tunisia, Central Mediterranean): A nursery area for sandbar shark, Carcharhinus plumbeus (Nardo, 1827) (Chondrichthyes: Carcharhinidae). Annales, Series Historia Naturalis. 2005;**15**:187-194

[15] Saidi B. Les requins du golfe de Gabès: Diversité et écobiologie de trois espèces à importance économique Carcharhinus plumbeus, Mustelus mustelus & M. punctulatus. [Thèse de doctorat en Sciences biologiques]. Université de Sfax, Tunisie. 2008. p. 192

[16] Enajjar S. Diversité des Rajiformes et etude eco–biologique de Rhinobatos rhinobatos et Glaucostegus cemiculus (Famille des Rhinobatidae) du Golfe de Gabès (Tunisie). [Thèse de doctorat en Sciences biologiques]. Faculté des Sciences de Sfax, Tunisie. 2009. p. 173

[17] Enajjar S, Saidi B, bradai MN. The Gulf of Gabes (Central Mediterranean Sea): A nursery area for sharks and batoids (Chondrichthyes: Elasmobranchii). Cahiers de Biologie Marine. 2015;**56**:143-150

[18] UNEP/MAP-SPA/RAC, AAO/ BirdLife in Tunisia, INSTM, DGPAq. In: Marouane B, Mourad BAM, Nejmeddine BM, Kamel EH, Samira E, Wasim G, et al., editors. Rapport technique, Résultats de la phase 2 des activités du programme d'observation des prises accessoires en Tunisie (novembre 2020- Janvier 2022) dans les GSA 12, 13 et 14. SPA/RAC, MedBycatch Project; 2022 (in press)